T0210638

Resistive Random Access Memory (RRAM)

From Devices to Array Architectures

Synthesis Lectures on Emerging Engineering Technology

Kris Iniewski, *Redlen Technologies, Inc.*

Resistive Random Access Memory (RRAM): From Devices to Array Architectures
Shimeng Yu
March 2016

The Digital Revolution
Bob Merritt
February 2016

Compound Semiconductor Materials and Devices
Zhaojun Liu, Tongde Huang, Qiang Li, Xing Lu, Xinbo Zou
February 2016

New Prospects of Integrating Low Substrate Temperatures with Scaling-Sustained Device Architectural Innovation
Nabil Shovon Ashraf, Shawon Alam, Mohaiminul Alam
February 2015

Advances in Reflectometric Sensing for Industrial Applications
Andrea Cataldo, Egidio De Benedetto, and Giuseppe Cannazza
January 2016

Sustaining Moore's Law: Uncertainty Leading to a Certainty of IoT Revolution
Apek Mulay
September 2015

Resistive Random Access Memory (RRAM): From Devices to Array Architectures
Shimeng Yu

ISBN: 978-3-031-00902-0 print
ISBN: 978-3-031-02030-8 ebook

DOI 10.1007/978-3-031-02030-8

A Publication in the Springer series
SYNTHESIS LECTURES ON EMERGING ENGINEERING TECHNOLOGIES, #6
Series Editors: Kris Iniewski, Redlen Technologies, Inc.

Series ISSN 2381-1412 Print   2381-1439 Electronic

Resistive Random Access Memory (RRAM)

From Devices to Array Architectures

Shimeng Yu
Arizona State University

SYNTHESIS LECTURES ON EMERGING ENGINEERING TECHNOLOGIES
#6

ABSTRACT

RRAM technology has made significant progress in the past decade as a competitive candidate for the next generation non-volatile memory (NVM). This lecture is a comprehensive tutorial of metal oxide-based RRAM technology from device fabrication to array architecture design. State-of-the-art RRAM device performances, characterization, and modeling techniques are summarized, and the design considerations of the RRAM integration to large-scale array with peripheral circuits are discussed. Chapter 2 introduces the RRAM device fabrication techniques and methods to eliminate the forming process, and will show its scalability down to sub-10 nm regime. Then the device performances such as programming speed, variability control, and multi-level operation are presented, and finally the reliability issues such as cycling endurance and data retention are discussed. Chapter 3 discusses the RRAM physical mechanism, and the materials characterization techniques to observe the conductive filaments and the electrical characterization techniques to study the electronic conduction processes. It also presents the numerical device modeling techniques for simulating the evolution of the conductive filaments as well as the compact device modeling techniques for circuit-level design. Chapter 4 discusses the two common RRAM array architectures for large-scale integration: one-transistor-one-resistor (1T1R) and cross-point architecture with selector. The write/read schemes are presented and the peripheral circuitry design considerations are discussed. Finally, a 3D integration approach is introduced for building ultra-high density RRAM array. Chapter 5 is a brief summary and will give an outlook for RRAM's potential novel applications beyond the NVM applications.

KEYWORDS

RRAM, ReRAM, resistive switching, non-volatile memory

Contents

CHAPTER 1

Introduction to RRAM Technology

1.1 OVERVIEW OF EMERGING MEMORY TECHNOLOGIES

The functionality and performance of today's computing system are increasingly dependent on the characteristics of the memory sub-system. The memory sub-system has a well-known hierarchy: From top layers to bottom layers, SRAM, DRAM, and FLASH are the mainstream memory technologies serving as cache, main memory, and solid-state-drive (SSD), respectively. Moving up the hierarchy, the memory write/read latency decreases. Moving down the hierarchy, the memory capacity increases. All these mainstream memory technologies are based on the charge storage mechanism: SRAM stores the charge at the storage nodes of the cross-coupled inverters, DRAM stores the charge at the cell capacitor, and FLASH stores the charge at the floating gate of the transistor. All these charge-based memories are facing challenges to be scaled down to 10 nm node or beyond, due to the easy loss of the stored charge at nanoscale, resulting in the degradation of the performance, reliability, and noise margin, etc. In this context, emerging memory technologies that are non-charge based are actively under research and development in the industry, with the hope to revolutionize the memory hierarchy [1].

The ideal characteristics for a memory device include fast write/read speed (<ns), low operation voltage (<1 V), low energy consumption (~fJ/bit for write/read), long data retention time (>10 years), long write/read cycling endurance (>10^{17} cycles), and excellent scalability (<10 nm). Nevertheless, it is almost impossible to satisfy these ideal characteristics in a single "universal" memory device. Several emerging non-volatile memory (NVM) technologies have been pursued toward to achieving part of these ideal characteristics. The emerging NVM candidates are spin-transfer-torque magnetoresistive random access memory (STT-MRAM) [2], phase change random access memory (PCRAM) [3], and resistive random access memory (RRAM) [4]. These emerging NVM technologies share some common features: they are non-volatile two-terminal devices, and they differentiate their states by the switching between a high resistance state (HRS, or off-state) and a low resistance state (LRS, or on-state). The transition between the two states can be triggered by electrical inputs. However, the detailed switching physics is quite different for different memories: STT-MRAM relies on difference in resistance between the parallel configuration (corresponding to LRS) and the anti-parallel configuration (corresponding to HRS) of two ferromagnetic layers separated by a thin tunneling insulator layer; PCRAM relies on chalcogenide material to switch between the crystalline phase (corresponding to LRS) and the amorphous phase (corresponding to HRS); and RRAM relies on the formation (corresponding to LRS) and the rupture (corresponding to HRS) of conductive filaments in the insulator between two electrodes. Due to the different underlying physics, the device characteristics are also different among different emerging NVM technologies. Table 1.1 compares the typical device characteristics between the emerging memory technologies and the mainstream memory technologies. It should be pointed out that different emerging NVM devices may have different application spaces due to their unique characteristics. As is seen from Table 1.1, compared to SRAM, STT-MRAM has an advantage of a smaller cell area, while STT-MRAM has maintained low programming voltage, fast write/read speed, and long endurance, thus STT-MRAM is attractive for embedded memories on chip, e.g., the SRAM replacement in the last-level cache [5]. Compared to FLASH, RRAM is attractive due to its lower programming voltage and faster write/read speed, thus the primary target of RRAM is to replace the NOR FLASH for code storage and more ambitiously to replace NAND FLASH as data storage [6]. Besides replacing the existing technologies, the emerging NVM technologies hold the potential to revolutionize today's memory hierarchy by adding more levels in the hierarchy, e.g., creating a storage class memory level between the main memory and the storage memory [7]. In addition, a hybrid system with emerging memories and mainstream memories is also attractive, e.g., using RRAM as the cache for NAND FLASH [8].

Table 1.1: Device characteristics of mainstream and emerging memory technologies							
	Mainstream Memories				Emerging Memories		
	SRAM	DRAM	FLASH		STT-MRAM	PCRAM	RRAM
			NOR	NAND			
Cell Area	>100F^2	6F^2	10F^2	<4F^2 (3D)	6~20F^2	4~20F^2	<4F^2 if 3D
Multi-bit	1	1	2	3	1	2	2
Voltage	<1V	<1V	>10V	>10V	<2V	<3V	<3V
Read Time	~1ns	~10ns	~50ns	~10µs	<10ns	<10ns	<10ns
Write Time	~1ns	~10ns	10µs–1ms	100µs–1ms	<5ns	~50ns	<10ns
Retention	N/A	~64ms	>10y	>10y	>10y	>10y	>10y
Endurance	>1E16	>1E16	>1E5	>1E4	>1E15	>1E9	>1E6~1E12
Write Energy (J/bit)	~fJ	~10fJ	100pJ	~10fJ	~0.1pJ	~10pJ	~0.1 pJ
F: feature size of the lithography, and the energy estimation is on the cell-level (not the array-level)							

1.2   RRAM BASICS

The resistive switching phenomenon where the resistance of insulators such as metal oxides changes when a large voltage is applied was firstly reported in the 1960s [9]. The recent revival on the resistive switching can be traced back to the discovery of hysteresis I-V characteristics in perovskite oxides such as $Pr_{0.7}Ca_{0.3}MnO_3$ [10], $SrZrO_3$ [11], $SrTiO_3$ [12], etc., in the late 1990s and the early 2000s. Since Samsung demonstrated NiO RRAM array integrated with the 180 nm silicon CMOS technology in 2004 [13], research activities have been blooming with demonstrations of resistive switching in various binary oxides[1] such as NiO [14], TiOx [15], CuO_x [16], ZrO_x [17], ZnO_x [18], HfO_x [19], TaO_x [20], AlO_x [21], etc., because of the simplicity of the materials and good compatibility with silicon CMOS fabrication process. Later in 2008, HP Labs made the connection of the resistive devices to the theoretical concept of memristor [22].[2]

Generally speaking, there are two types of RRAM. The first type is based on the conductive filaments consisting of oxygen vacancies, which is typically referred to as oxide-based RRAM; the second type is based on the conductive filaments consisting of metal atoms, which is also called conductive-bridge RAM (CBRAM). CBRAM relies on the fast-diffusive Ag or Cu ions migration into the oxide (or chalcogenide) to form a conductive bridge. Despite different underlying switching physics, these two types share a lot of common device characteristics and the array architecture design considerations are very similar. In this lecture, we focus on the first type: oxygen vacancy based oxide RRAM.[3] In literature, there are a few comprehensive reviews on the oxide RRAM [23, 24, 25, 4]. For the CBRAM, refer to the review [26].

So far, tens of binary oxides have been found to exhibit resistive switching behavior. Most of them are transition metal oxides, and some are lanthanide series metal oxides. The materials for the resistive switching oxide layer and the electrodes reported in literature are summarized in Table 1.2. Besides metals, conductive nitrides, e.g., TiN, TaN, are also commonly used as electrode materials.

---

[1] These binary oxides that show resistive switching are often non-stoichiometric, thus subscript "x" is used for the oxygen composition in this book.

[2] To unify the terminology and emphasize the technology development, RRAM is used instead of memristor in this book.

[3] If not specifically noted, the term "RRAM" refers to the binary oxide memory involving oxygen vacancies in this book.

Table 1.2: Summary of the materials that have been used for binary oxide RRAM reported in literature . Metals of the corresponding binary oxides used for the resistive switching layer are colored in yellow. Metals used for the electrodes are colored in blue. Used with permission from [4].

The Periodic Table of the Elements

corresponding binary oxide that exhibits bistable resistance switching

metal that is used for electrode

1 H																1 H	2 He
3 Li	4 Be											5 B	6 C	7 N	8 O	9 F	10 Ne
11 Na	12 Mg											13 Al	14 Si	15 P	16 S	17 Cl	18 Ar
19 K	20 Ca	21 Sc	22 Ti	23 V	24 Cr	25 Mn	26 Fe	27 Co	28 Ni	29 Cu	30 Zn	31 Ga	32 Ge	33 As	34 Se	35 Br	36 Kr
37 Rb	38 Sr	39 Y	40 Zr	41 Nb	42 Mo	43 Tc	44 Ru	45 Rh	46 Pd	47 Ag	48 Cd	49 In	50 Sn	51 Sb	52 Te	53 I	54 Xe
55 Cs	56 Ba	57 La	72 Hf	73 Ta	74 W	75 Re	76 Os	77 Ir	78 Pt	79 Au	80 Hg	81 Tl	82 Pb	83 Bi	84 Po	85 At	86 Rn
87 Fr	88 Ra	89 Ac	104 Rf	105 Db	106 Sg	107 Bh	108 Hs	109 Mt	110	111	112		114		116		118

58 Ce	59 Pr	60 Nd	61 Pm	62 Sm	63 Eu	64 Gd	65 Tb	66 Dy	67 Ho	68 Er	69 Tm	70 Yb	71 Lu
90 Th	91 Pa	92 U	93 Np	94 Pu	95 Am	96 Cm	97 Bk	98 Cf	99 Es	100 Fm	101 Md	102 No	103 Lr

Before starting the discussion in this lecture, we first introduce some basic concepts and terminologies about RRAM. Figure 1.1 (a) shows the typical metal-insulator-metal device structure of RRAM: a thin oxide layer sandwiched between two electrodes. The switching event from HRS to LRS is called the "set" process. Conversely, the switching event from LRS to HRS is called the "reset" process. Usually for the fresh samples, the initial resistance is very high and a large voltage is needed for the first cycle to trigger the switching behaviors for the subsequent cycles. This is called the "forming" process. The switching modes of RRAM can be broadly classified into two switching modes: unipolar and bipolar. Figure 1.1 (b) and (c) show a sketch of the I-V characteristics for the two switching modes. Unipolar switching means the switching direction depends on the amplitude of the applied voltage but not on the polarity of the applied voltage. Thus set/reset can occur at the same polarity. If the unipolar switching can symmetrically occur at both positive and negative voltages, it is also referred to as a nonpolar switching mode. Bipolar switching means the switching direction depends on the polarity of the applied voltage. Thus set can only occur at one polarity and reset can only occur at the reversed polarity. For either switching mode, to avoid a permanent dielectric breakdown in the forming/set process, it is recommended to enforce a compliance current, which is usually provided by the semiconductor parameter analyzer during off-chip testing, or more practically, by a cell selection device (transistor, diode, or a series resistor) on-chip. To read the data from the cell, a small read voltage that does not affect the memory state is applied to detect whether the cell is in HRS or LRS.

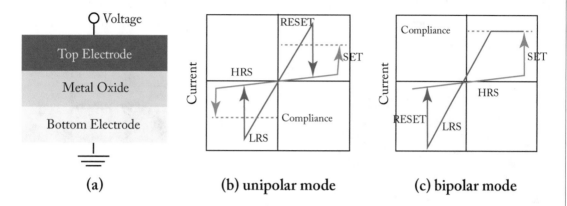

Figure 1.1: (a) Schematic of metal-insulator-metal structure for oxide RRAM, and schematic of RRAM's I-V curves, showing two modes of operation: (b) unipolar and (c) bipolar. Adapted from [4].

It is observed that the electrode materials have a significant impact on the switching modes of the oxide RRAM. Even with the same oxide material but with different electrode materials, the switching modes can be different. Therefore, it is inferred that the switching mode is not an intrinsic property of the oxide itself but a property of both the oxide materials and the electrode/oxide interfaces. In most cases, the unipolar mode is obtained with a noble metal such as Pt as both top and bottom electrodes. With one of the electrodes replaced by oxidizable materials such as Ti or TiN, the bipolar mode is obtained. For the bipolar switching, it was suggested that the reversed field is required for a successful reset because an interfacial oxygen barrier (e.g., TiON) exists [27]. If both electrodes are oxidizable, there should be some asymmetry in the oxygen gettering capability for the bipolar switching. A typical structure is TiN/metal/oxide/TiN,[4] e.g., TiN/Ti/HfO$_x$/TiN [19], TiN/Hf/HfO$_x$/TiN [28], where the metal capping layer functions as an oxygen gettering layer. Generally, the unipolar switching needs a higher reset current than the bipolar switching, and shows a larger variability as well. Therefore, today RRAM research and development focus more on the bipolar switching, and we will focus our discussions on the bipolar mode in this book.[5]

To further understand how the RRAM device works as NVM, the device characteristics of HfO$_x$-based RRAM [19, 29] from Industrial Technology Research Institute (ITRI) at Taiwan are presented here as an example. Figure 1.2 (a) shows the transmission electron microscopy (TEM) image of the concave structure device of TiN/Ti/HfO$_x$/TiN stack with 30 nm cell size. Figure 1.2

---

[4]   If not specifically noted, the stack sequence is from the top electrode to the bottom electrode in this book.
[5]   If not specifically noted, the switching mode is bipolar switching in this book.

(b) shows the typical I-V curve of this RRAM cell. A 200 μA set compliance current is enforced, and the device exhibits a bipolar switching. Figure 1.2 (c) shows the programming cycling endurance characteristics. The set/reset programming condition is +1.5 V/-1.4 V pulse with 500 μs width, and after 10^6 switching cycles the resistance on/off ratio is still larger than 100. Figure 1.2 (d) shows the data retention testing results. The device is baked at 150°C, and i a 10-year lifetime is expected using a simple linear extrapolation.

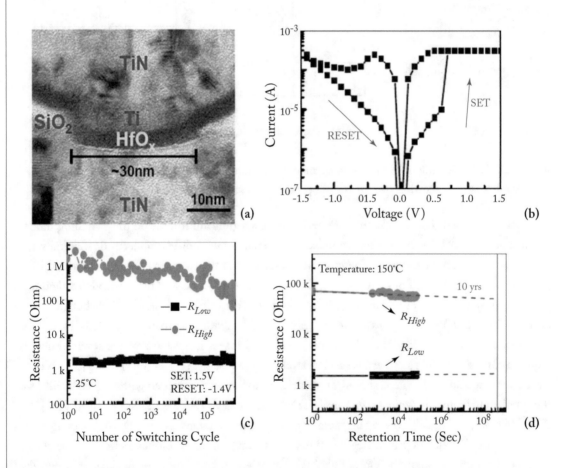

Figure 1.2: (a) Transmission electron microscopy (TEM) image of the concave structure device of ITRI's TiN/Ti/HfOx/TiN stack; (b) typical I-V curve of the device with the cell size of 30 nm; (c) programming endurance cycling by 500 μs pulse: successive 10^6 switching cycles is achieved; (d) retention testing at 150 °C: a 10-year lifetime is expected. Adapted from [19, 29].

1.3   RECENT RESEARCH AND DEVELOPMENT OF RRAM TECHNOLOGY

The development of oxide RRAM has progressed rapidly in the past decade. In particular, binary oxides that use materials that are compatible to the silicon CMOS fabrication process have seen intense research and development in the industry. The early RRAM devices in the mid-2000s had large device area ($>>\mu m^2$), large programming current ($\sim mA$), long programming time ($>\mu s$), low endurance ($<10^3$ cycles), and required a large forming voltage (~ 10 V). Today, many of these deficiencies have been overcome. Device sizes down to 10 nm or below have been demonstrated [28, 30], programming current is now in the order of tens of μA or a few μA, programming speed is in the order of tens of ns or a few ns, programming endurance cycles are typically larger than 10^6 with a record up to 10^{12} [31], retention time is $>3,000$ h at $150°C$ and extrapolated to be more than 10 years at $85°C$ [20], and the forming process can be eliminated by shrinking the oxide thickness [32] or other oxide stack engineering strategies. Most of these good characteristics were reported in HfO_x or TaO_x systems. Demonstrations of 2-bit and 3-bit multi-level operation have also been made [33, 34]. Chip-level RRAM array macro from 4 Mb to 32 Gb capacity with peripheral circuitry have been demonstrated by industry as well [35, 36, 37], showing that RRAM is a viable NVM technology for practical applications.

This book is organized as follows. Chapter 2 will discuss the RRAM device fabrication techniques and methods to eliminate the forming process, and will show its scalability down to sub-10 nm regime. Then the device performances such as programming speed, variability control, and multi-level operation will be presented, and finally the reliability issues such as cycling endurance and data retention will be discussed. Chapter 3 will discuss the RRAM physical mechanism, the materials characterization techniques to observe the conductive filaments, and the electrical characterization techniques to study the electronic conduction processes. It will also present the numerical device modeling techniques for simulating the evolution of the conductive filaments as well as the compact device modeling techniques for circuit-level design. Chapter 4 will discuss the two common RRAM array architectures for large-scale integration: one-transistor-one-resistor (1T1R) and cross-point architecture with selector. The write/read schemes are presented and the peripheral circuitry design considerations are discussed. Finally, a 3D integration approach is introduced for building ultra-high density RRAM array. Chapter 5 is a brief summary and will give an outlook for RRAM's potential novel applications beyond the NVM applications.

CHAPTER 2

RRAM Device Fabrication and Performances

2.1 DEVICE FABRICATION: FORMING-FREE AND SCALABILITY

The metal oxide RRAM device fabrication uses mainly conventional semiconductor fabrication tools, and it is compatible with the silicon CMOS back-end-of-line (BEOL) process where low-temperature (<400°C) is required. To deposit the resistive switching oxide layer, two typical methods are used: (1) physical vapor deposition (PVD), i.e., sputter from metal targets followed by annealing in oxygen ambient or reactive sputtering in oxygen ambient, and the sputter temperature can be as low as room temperature; (2) atomic layer deposition (ALD) from metal-organic precursor in water or ozone ambient, and the typical ALD temperature is around 200°C. In this chapter, we will use the European microelectronics research institute IMEC's TiN/Hf/HfO$_x$/TiN devices [28] to illustrate the RRAM cell design and its impact on the forming behavior. The IMEC devices are based on 65 nm silicon CMOS process and built on top of a transistor's drain contact via. Figure 2.1 (a) shows the flow for RRAM BEOL integration: After the transistor is done with front-end-of-line (FEOL) process for the transistor, bottom electrode TiN is deposited by PVD and then followed by the chemical mechanical planarization (CMP). HfO$_x$ resistive switching layer is deposited by ALD, and then Hf capping layer and TiN top electrode is deposited by PVD. The Hf capping layer functions as the oxygen-gettering layer that attracts the oxygen from the HfO$_x$ layer and make it sub-stoichiometry (x<2). Finally, the top electrode is patterned and followed by the top-level passivation. Figure 2.1 (b)–(d) show the TEM images of the fabricated cell down to an active size of 10 nm×10 nm.

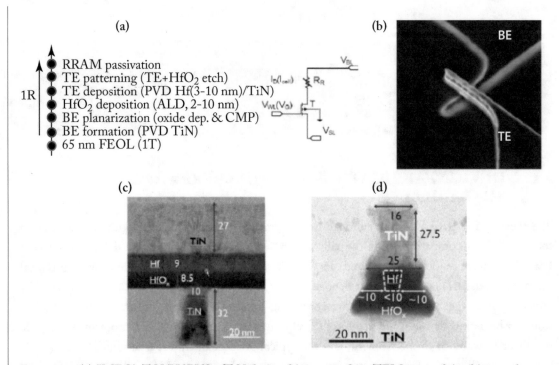

Figure 2.1: (a) IMEC's TiN/Hf/HfOx/TiN device fabrication flow. TEM view of the fabricated device in bird-eye view (b) and two cross-sectional views (c)–(d). The device active cell area is scaled down to 10 nm×10 nm. Adapted from [28].

There are two key geometrical parameters (HfO$_x$ thickness and Hf capping layer thickness) that affect the device characteristics (especially the forming behaviors). In addition, the forming voltage is found to be a strong function of the device area. Figure 2.2 (a) shows that the forming voltage undesirably increases when scaling the device area. This is because the forming process is similar as a dielectric soft-breakdown, and the percolation theory [38] suggests that the breakdown voltage depends on the number of defects rather than the defect density in the dielectric layer. With the scaling, the number of as-fabricated defects (i.e., oxygen vacancies) decreases and the probability to form a percolation conductive path reduces, therefore a higher voltage is needed to create more defects and a conductive filament. Figure 2.2 (a) also shows that the poly-crystalline HfO$_x$ has a lower forming voltage than the amorphous HfO$_x$ probably due to the leakage through the grain boundaries. Typically, HfO$_x$ as deposited by ALD is amorphous, and a 600°C annealing can partially crystalize the thin film. Nevertheless, the poly-crystalline HfO$_x$ RRAM's performance is not as good as the amorphous one. To reduce the forming voltage, reducing the HfO$_x$ layer thickness is a better approach. For a 10 nm×10 nm cell, the forming voltage reduces from 5.3 V for 10 nm HfO$_x$ layer to 2.3 V for 5 nm HfO$_x$ layer, and eventually forming-free is achieved for 2 nm HfO$_x$ layer (both forming voltage and set voltage in the subsequent cycles <1 V). At such extremely thin thickness, a precise control of the electrode surface roughness is required to prevent short-circuit due to the discontinuity of the oxide thin film; thus the ALD is a preferred fabrication method due to its uniform coverage capability. In addition, a thicker metal capping layer can reduce the forming voltage as shown in Figure 2.2 (b), because it may attract more oxygen from the oxide layer and make the oxide layer more oxygen deficient. However, the methods for fabricating a forming-free device such as reducing oxide thickness and increasing the initial defect density by thicker metal capping layer may also severely decrease the resistivity of the oxide layer and sacrifice the memory on/off ratio.

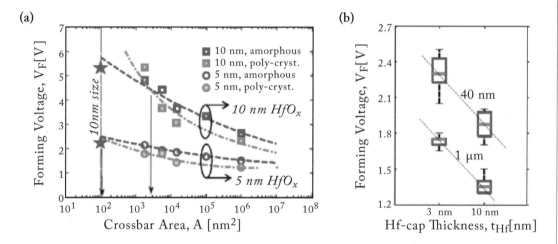

Figure 2.2: Forming voltage trend of IMEC's HfO$_x$-based RRAM. (a) Forming voltage with the scaling of device area for a 10 nm and 5 nm amorphous and poly-crystalline HfO$_x$ layer; (b) forming voltage with the scaling of Hf metal capping layer thickness for a 1 µm and 40 nm cell size. Adapted from [28].

The scalability of RRAM is proved to be excellent, and sub-10 nm devices have been successfully fabricated. For example, an extremely small HfO$_x$-based RRAM with active size 1 nm×3 nm has been fabricated at the sidewall [30], exhibiting reasonably good performances such as on/off ratio (>100), endurance (>10^4 cycles), and retention (> 2×10^4 h at 250°C). Figure 2.3 shows the scaling trend of device parameters such as set/reset voltages, and on/off ratio from 1 µm×1 µm down to 10 nm×10 nm for IMEC's HfO$_x$-based RRAM. It is seen that all these device parameters show very weak dependence on the cell area, unlike the forming voltage discussed earlier, suggesting a filamentary switching mechanism. Once the RRAM devices are formed, in the subsequent cycles, the switching occurs at a localized region that is much smaller than the actual device size. The lateral diameter of the conductive filament can range from tens of nm down to a few nm, depending on the current in the LRS. Along the filament direction, the active region for the switching is also confined (probably within one or two nm), as the filament is only partially ruptured.

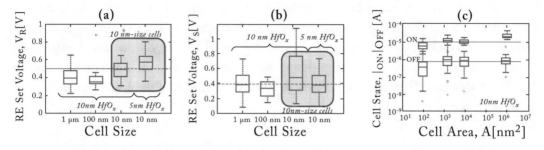

Figure 2.3: Scaling trend of device parameters such as (a) reset voltage, (b) set voltage, and (c) on/off ratio from 1 μm×1 μm down to 10 nm×10 nm for IMEC's HfOx-based RRAM. Adapted from [28].

2.2   DEVICE PERFORMANCES

In this section, we will discuss the device performances, including the programming speed, the variability, and the multi-level operation. RRAM's programming speed can be as fast as a few ns. However, the programming speed is a strong function of the programming voltage. Figure 2.4 shows the relationship between set/reset voltage and set/reset pulse duration for IMEC's HfO$_x$-based RRAM. Roughly speaking, increasing about 0.25 V and 0.5 V of programming voltage will increase the programming speed by one order of magnitude for 1 μm cell and 10 nm cell, respectively. This exponential voltage-time relationship is attributed to the energy barrier lowering effect of the oxygen vacancies' generation and migration physics [39]. Although RRAM's programming speed can be generally improved to sub-10 ns regime by increasing the programming voltage, cautions should be taken to prevent the damages to the cell by using a large voltage. The programming speed at sub-ns regime is pretty much unexplored so far, due to the difficulty of off-chip measurement, e.g., the parasitic capacitances of pads and cables can significantly distort the waveforms at sub-ns regime. Nevertheless, 300 ps programming was reported in ITRI's HfO$_x$-based RRAM, and 300 ps is actually limited by the speed of instrument [40]. The intrinsic RRAM switching speed limit may be even faster, because the single-event-upset that can switch the RRAM states is observed from the heavy ion strike radiation experiments, and the duration of the ion-induced photocurrent transient is typically tens of ps [41].

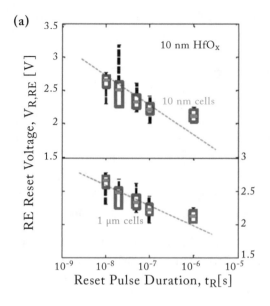
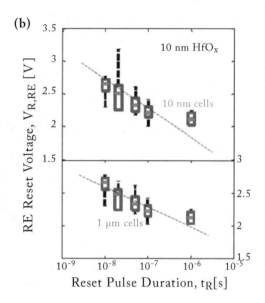

Figure 2.4: Relationship between set/reset voltage and set/reset pulse duration for IMEC's HfO$_x$-based RRAM. Adapted from [28].

Variability of device characteristics is a major barrier to large-scale manufacturing of RRAM. Significant parameter fluctuations exist in terms of variations of the set/reset voltages as well as the resistances in HRS and in LRS. The variations include temporal (cycle to cycle) variations and spatial (device to device) variations. The spatial variations may be improved with the precise manufacturing control of uniformity across wafers. However, the temporal variations seem an intrinsic property of the RRAM switching dynamics caused by the stochastic nature of the oxygen vacancies generation and migration processes [42]. Typically, the HRS resistance variation is more remarkable than the LRS resistance variation. The LRS resistance variation comes from the variation of the diameter of the conductive filament or the number of the conductive filaments, while the HRS resistance variation comes from the variation of the ruptured filament distance, thus any small variation of the gap distance may be magnified to be a variation of tunneling current in HRS. It is found that the tail bits of the HRS distribution may be associated with the oxygen vacancies left over inside the ruptured filament region [42].

RRAM can be used as multi-level cell (MLC) to increase its density. RRAM modulates the resistance states into multi-levels to realize the MLC operation. There are two ways to modulate the RRAM resistance states, either by controlling the set compliance current or by controlling the reset voltage. As shown in ITRI's HfO$_x$-based RRAM [19] in Figure 2.5, the LRS resistance can be changed by the set compliance current possibly due to the modulation of the diameter of the conductive filament or the number of filaments, while the HRS resistance can be controlled by the reset voltage possibly due to the modulation of the ruptured filament distance. These multi-level resistance states can maintain their states at 85°C for 10 years by a simple linear extrapolation.

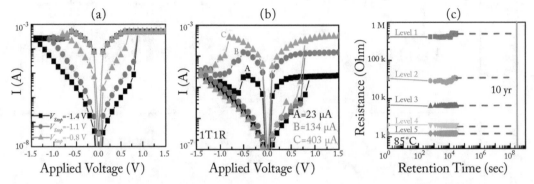

Figure 2.5: Multi-level characteristics for ITRI's HfO$_x$-based RRAM: (a) multi-level HRS is obtained by controlling reset stop voltage and (b) multi-level LRS is obtained by controlling the set compliance current. (c) The retention of multi-level resistance states extrapolated to 10 years at 85°C. Adapted from [19].

As described earlier, the variability of the RRAM resistances is significant. MLC operation requires a very tight control of the resistance distribution because the memory window should be distinguishable between levels. In practice, the write-verify programming scheme is used to tighten the resistance distribution for MLC. Figure 2.6 shows the effect of successively ramping up the set compliance current (determined by the gate voltage for the series transistor) to the desired level. As the compliance current is ramped up, the resistance decreases further. If the resistance over-set to a lower resistance than the target level, a reset operation is performed and the compliance current ramping is re-attempted to achieve the target resistance. Nevertheless, the write-verify programming scheme will sacrifice the programming speed. In ITRI's 4 Mb HfO$_x$-based RRAM prototype chip design [35], single-level cell (SLC) can achieve 7.2 ns programming speed, while MLC (2 bit/cell) needs 160 ns to perform the write-verify scheme mentioned above. The thermal and voltage-stress stability of the resistance in each state is also important for the MLC operation. In ITRI's HfO$_x$-based RRAM, a 4-level cell remains stable for 3×10^4 s above 85 °C, and good immunity was demonstrated for total read stress of 2×10^4 s (20 ms read for 10^6 cycles). So far, the largest number of 8-level MLC operation (3 bit/cell) was demonstrated in WOx based RRAM [33].

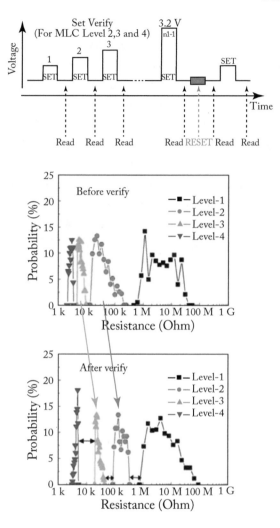

Figure 2.6: Write-verify prorgamming scheme for MLC operation in ITRI's HfO$_x$-based RRAM. The verification scheme consists of ramping up the gate voltage (compliance current), but applying a reset and repeating re-ramping in order to avoid excessively low resistance. The desired outcome after verification is shown at the bottom. Adapted from [35].

2.3   DEVICE RELIABILITY

Reliability of RRAM has two aspects: cycling endurance and data retention. Generally, the cycling endurance refers to how many cycles that the device can be programmed. There are different failure modes of the cycling endurance depending on the programming conditions [43]. Figure 2.7 shows an example of cycling endurance of IMEC's HfO$_x$-based RRAM under different programming conditions. In these experiments, the gate voltage or word line (WL) voltage of the series transistor is varied to change the set compliance current while other programming parameters are fixed. A weak set condition (smaller WL voltage) tends to result in a set failure (the device is stuck at HRS and cannot set after 10^6 cycles), and a strong set condition (larger WL voltage) tends to result in a reset failure (the device is stuck at LRS and cannot reset after 10^6 cycles). Therefore, the relative strength of the set and reset condition determines the failure modes. Similar observations are found in varying the reset voltage amplitude while fixing other programming parameters. Therefore, a balanced set/reset condition is important to improve the cycling endurance. Figure 2.8 shows cycling endurance for IMEC's HfO$_x$-based RRAM with optimally balanced set/reset programming conditions (set: WL =1 V, BL = 1.8 V, and width=5 ns; reset: WL =3 V, SL = 1.8 V, and width=10 ns). A stable ~15× on/off ratio could be achieved for 10^{10} cycles. So far, the largest number of cycling endurance (10^{12} cycles) was demonstrated in TaO$_x$-based RRAM [31].

	SET	WL = *0.9 V*	WL = *1 V*	WL = *1.2 V*	WL = *1.4 V*
RESET		100 ns	100 ns	100 ns	100 ns
SL = 1.8 V, 10 ns		(a)	(b)	(c)	(d)

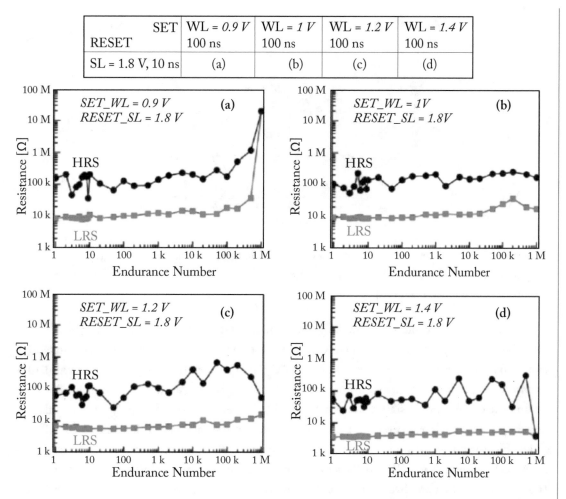

Figure 2.7: Cycling endurance of IMEC's HfOx-based RRAM under different programming conditions. From (a) to (d), the gate voltage or word line (WL) voltage of the series transistor is varied to increase the set compliance current while other programming parameters are fixed. A weak set condition (smaller WL voltage) tends to result in a set failure, and a strong set condition (larger WL voltage) tends to result in a reset failure. Adapted from [43].

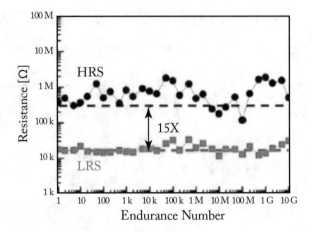

Figure 2.8: Cycling endurance for IMEC's HfO$_x$-based RRAM with optimally balanced set/reset programming conditions (set: WL =1 V, BL = 1.8 V, and width=5 ns; reset: WL =3 V, SL = 1.8 V, and width=10 ns). Adapted from [43].

Data retention refers to how long the memory states can be maintained. Typically, data retention time longer than 10 years (~3×10^8 s) is expected for NVM applications. This retention should be maintained at thermal stress up to 85°C (i.e., the operating temperature on chip). One common simple linear extrapolation method in literature is to bake the devices at a high temperature (e.g., on a probe station), and monitor the device's resistance by applying read pulses at certain time intervals, e.g., every 1 s, and extrapolate the resistance evolution line to the 10-year point. However, this method, while easy to be implemented in an industrial test setting, has its limitations. Although the RRAM devices can maintain the resistance window over 10^4 s or 10^5 s (the time period that is convenient for testing), it cannot be guaranteed that the resistance window still exists after 10^6 or 10^7 s if the resistance window collapses abruptly rather than gradually. Also in this method, the read voltage stress is applied to the cell during the retention test. To minimize the effect of the read voltage stress, another often used method is to bake the device at elevated temperatures (e.g., in an oven) for an extended period and then read out the resistances at specific times (after cool-down), e.g., 24 h, 100 h, and so on. The most accurate method is the temperature-accelerated testing by varying the temperature of the baking: record the time-to-failure at each temperature, draw the Arrhenius (1/kT) plot to extract the activation energy, and then extrapolate down to the operating temperature. In this method, one has to wait until the failure occurs, thus it is more time-consuming. Figure 2.9 shows an example of temperature-accelerated retention test for IMEC's HfO$_x$-based RRAM [44]. Different set compliance current (100 μA and 10 μA) were used to achieve two LRS levels. A lower compliance current results in worse LRS data retention because a weaker filament tends to rupture easily under high temperature. Therefore, a trade-off exists between the low power operation and the long data retention. Using the Arrhenius (1/kT) plot, an activation energy (E_a~1.5 eV) was extracted for IMEC's HfO$_x$-based RRAM (with compliance current = 10 μA). In IMEC's devices, the HRS degradation has a similar trend as the LRS degradation: both the LRS and HRS resistances increase over the baking time, and similar E_a was extracted for the HRS degradation, suggesting the same underlying physics. In this case, the LRS degradation is the limiting factor for the data retention as the LRS resistance may increase above the reference between LRS and HRS. To improve the LRS data retention, post-fabrication annealing at 400°C was proposed to increase the oxygen content in the metal capping layer, thereby forming an interfacial layer between the metal capping layer and the oxide layer that reduces the mobility of the oxygen vacancies [45]. In addition, the data retention characterization should be combined with the endurance characterization as the post-cycling data retention becomes worse as more cycles have been programmed to the cell. The IMEC's HfO$_x$-based RRAM after 10^6 programming cycles shows a remarkable degradation in LRS data retention compared to the fresh devices [46]. Extensive data retention testing has also been performed in Panasonic's TaO$_x$-based RRAM. Similarly, the LRS resistance tends to increase with the baking time. The activation energy extracted for TaO$_x$-based RRAM is about 1.2 eV [20]. It should be mentioned that realistic statistics on retention property can only be collected on large memory arrays and the tail bits of the failure time distribution becomes the limiting factor for the entire array. In Panasonic's 256 kb TaO$_x$-based RRAM array, no data retention failure happens after 1,000 h at 150°C, and even the tail bits can exceed more than 10 years at 85°C by 1/kT extrapolation [47].

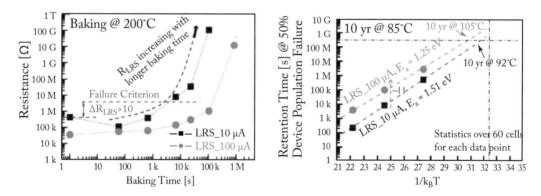

Figure 2.9: (a) Temperature-accelerated retention test for IMEC's HfO$_x$-based RRAM at 200°C (b) Arrhenius plot of retention time to failure at three temperature 250°C, 200°C, and 150°C. An activation energy (Ea~1.5 eV) was extracted for cells with compliance current = 10 μA. Adapted from [44].

CHAPTER 3

RRAM Characterization and Modeling

3.1 OVERVIEW OF RRAM PHYSICAL MECHANISM

The physical mechanism for the metal-oxide RRAM has been a complicated topic under debate for years. In general, the resistive switching is associated with the generation of oxygen vacancies (Vo) and migration of oxygen ions (O^{2-}) to form conductive filament or filaments between the two electrodes. This process is usually accompanied by electrochemical reactions, therefore it is also referred to as the redox (reduction/oxidation) effect [24]. In this lecture, we will focus on the filamentary switching mechanism, which is the prevailing theory for most of the metal-oxide RRAM. For the discussion on non-filamentary or interfacial barrier modulation mechanism, one can refer to [48].

Although the details of the resistive switching physics is still an active research area, here we aim to give a general physical picture for the filamentary switching mechanism, as shown in Figure 3.1. The forming process for the fresh samples is similar as a dielectric soft breakdown. Initially, the Vo density is low. Under the high electric field (>10 MV/cm), the oxygen atoms are knocked out of the lattice, and become O^{2-} drifting toward the anode and Vo are left in the oxide layer. O^{2-} are discharged as neutral non-lattice oxygen if the anode materials are noble metals or react with the oxidizable anode materials to form an interfacial oxide layer. Thus the electrode/oxide interface behaves like an "oxygen reservoir" [49]. Meanwhile, Vo in the bulk oxide (the localized deficiency of oxygen) leads to the formation of conductive filament (CF). Now the RRAM device switches to the LRS. Usually the as-deposited RRAM oxide thin films are amorphous or poly-crystalline, and the CFs are preferentially generated along the grain boundaries [50]. The roughness of the electrode/oxide interface may also concentrate the CF into the electric-field enhanced region. During the reset process, O^{2-} migrate back from the interface to the bulk oxide either to recombine with the Vo or to oxidize the metallic CF precipitates and thus partially rupture the CF. For the unipolar switching, Joule heating by the current thermally activates the O^{2-} diffusion, thus O^{2-} diffuse from the interface or the region around the CF due to the concentration gradient [51]. Usually the unipolar switching requires a relatively higher reset current to raise the local temperature around CF. For the bipolar switching, the interfacial layer may present a significant diffusion barrier and pure thermal diffusion is not sufficient, thus O^{2-} migration needs to be aided by the reverse electric field [27]. Nevertheless, in both cases the CF is partially ruptured, and a Vo-poor region forms and causes a tunneling gap for electrons. Now the RRAM device switches to the HRS. The residual CF with Vo-rich region is referred to as the "virtual electrode." In the next set process, the soft-breakdown occurs in the gap region and the CF reconnects both electrodes. Then such set/reset loop can repeat for many cycles.

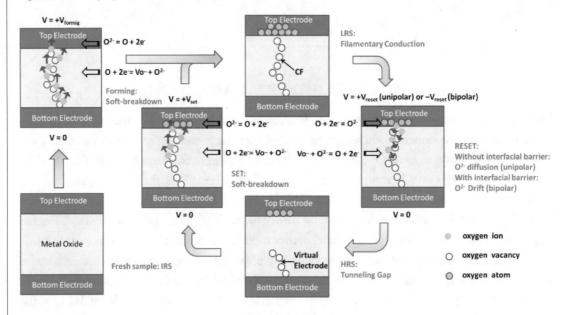

Figure 3.1: Schematic of the filamentary switching mechanism of the metal-oxide RRAM. Used with permission from [4].

3.2   MATERIALS AND ELECTRICAL CHARACTERIZATION

The conduction of LRS current in the metal oxide RRAM is usually filamentary. Conductive atomic force microscopy (C-AFM) is a useful method to observe the CFs. To observe the CFs intrinsically formed underneath the electrodes instead of artificially manipulated by C-AFM, a technique is developed by IMEC to remove the electrode materials after the normal set/reset operations on the electrodes of the devices [52]. To minimize the changes to the CFs during the sample preparation, the electrode layer is removed physically by means of the shear force exerted during repetitive high-pressure scans with an AFM tip on the electrode layer. Figure 3.2 shows the C-AFM image of IMEC's HfO_x-based RRAM devices in fresh state, LRS and HRS after the electrode removal, respectively. Firstly, the fresh cell does not show any leakage current. Secondly, the LRS cell shows a dominant filament with an observed diameter ranging between 30 and 50 nm. The current carried by these conductive paths is on the order of 1 nA. The conduction is fairly homogeneous within the conductive spot. Finally, the HRS cell shows a much reduced filament residual with a smaller diameter (5–10 nm), and the leakage current reduces to the pA range.

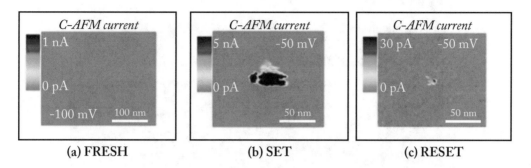

(a) FRESH   (b) SET   (c) RESET

Figure 3.2: Conductive atomic force microscopy (C-AFM) image of IMEC's HfO_x-based RRAM devices in (a) fresh state, (b) LRS, and (c) HRS after the electrode removal, respectively. Adapted from [52].

The first direct observation of CFs in the cross-section fashion was done in the TiO_x-based RRAM [53], in which the nanoscale (~10 nm in diameter) CF was seen by high-resolution transmission electron microscopy (HR-TEM). The nature of CFs is usually conjectured to be Vo in the metal-oxide RRAM devices. It is well known that Vo can act as an effective donor in n-type metal oxides. Taking HfO_x-based RRAM as an example, ab-initio calculation of the monoclinic and amorphous HfO_x electronic structure indicates that Vo can produce a defect state within the band gap [54]. Ordered Vo chain can form a transmission channel between two metal contacts. Experimentally, HR-TEM measurements [55] on the HfO_x-based RRAM show that the CF extends ~20 nm diameter with morphological changes and local atomic disorder. The electron energy loss spectroscopy (EELS) on oxygen K-edges spectrum reveals the presence of Vo associated with localized states in the band gap inside the filament region. However, the composition of CFs is not limited to Vo. Sometimes, the CFs can be metallic as well. Another study [56] using HR-TEM and EELS composition mapping shows an observation of a 5 nm to 15 nm wide filament dominated by metallic Hf in the HfO_x-based RRAM, as shown in Figure 3.3.

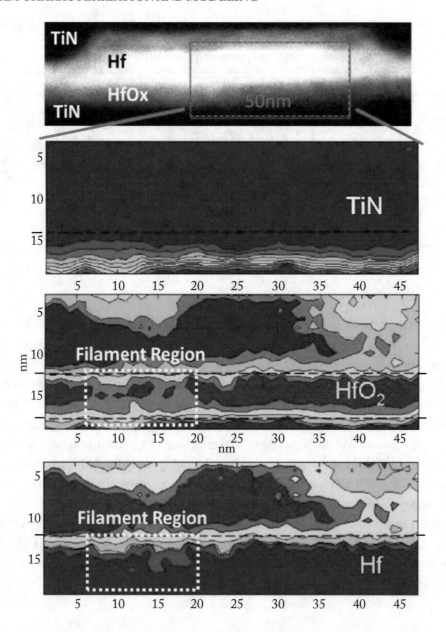

Figure 3.3: High-resolution transmission electron microscopy (HR-TEM) and electron energy loss spectroscopy (EELS) composition mapping of TiN, HfO_2 and Hf in the HfO_x-based RRAM. 5 nm to 15 nm wide filament dominated by metallic Hf is observed. Adapted from [56].

We believe that the nature of CFs depends on the density of Vo. If the density of Vo is low, then the CFs may show a semiconducting behavior as electrons are still in localized states in the band gap. If the density of Vo is sufficiently high, then the CFs may show a metallic behavior as electrons are now in extended states forming a sub-band in the band gap. A simple way to distinguish whether the CFs are metallic or semiconducting is to measure the temperature dependence of the resistance. If the resistance decreases with the decrease of temperature, the CFs are metallic and may consist of metal precipitates. On the contrary, if the resistance increases with the decrease of temperature, the CFs are semiconducting and may consist of Vo. In a low temperature measurement (down to 4 K) on the HfO_x-based RRAM [57], the normal LRS and HRS states (>10 kΩ) both show the semiconducting behavior, while the metallic behavior was observed for the failure state with extremely low resistance (<3 kΩ), suggesting that the excess Vo are generated in the failure state.

There are many efforts to fit the I-V characteristics to analyze the current conduction mechanism of the RRAM devices in literature. Most reports show a linear or an Ohmic relationship in LRS. However, the conduction characteristics in HRS can be fitted with various models: Poole-Frenkel emission ($\log(I/V) \sim V^{1/2}$), Schottky emission ($\log(I) \sim V^{1/2}$), the space charge limited current (SCLC) characteristic (the Ohmic region I~V, and the Child's square law region I~V^2). We think that a simple I-V fitting with the aforementioned established model may not be sufficient to ascertain the conduction mechanism in the metal oxide RRAM. In general, Figure 3.4 shows all the possibilities for electrons to transport from cathode to anode [58]: (1) Schottky emission: thermally activated electrons injected over the barrier into the conduction band; (2) Fowler-Nordheim (F-N) tunneling: electrons tunnel from the cathode into the conduction band, usually occurs at high field; (3) direct tunneling: electrons tunnel from cathode to anode directly, usually occurs when the oxide is thin enough (<3 nm). If the oxide has a substantial number of traps (e.g., Vo), trap-assisted-tunneling (TAT) contributes to additional conduction, including the following steps: (4) tunneling from cathode to traps; (5) emission from trap to conduction band, which is essentially the Poole-Frenkel emission; (6) F-N-like tunneling from trap to conduction band; (7) trap-to-trap hopping or tunneling, maybe in the form of Mott hopping when the electrons are in the localized states, or maybe in the form of metallic conduction when the electrons are in the extended states depending on the overlap of the electron wave function; and (8) tunneling from traps to anode. Whether any one particular process dominates is determined by its transition rate; electrons would seek the fastest transition (or least resistive) paths among all the possibilities. Therefore, various oxide RRAM devices may have different dominant conduction mechanisms depending on the dielectric properties (band gap or trap energy level, etc.), the fabrication process conditions (annealing temperature, annealing ambient, etc.), and the properties of the interface between the oxides and the electrodes (interfacial barrier height). The I-V relationship at the low bias regime is mainly determined by the electron conduction process for a given configuration of the CF, while at the high bias regime, the motion of atoms (such as Vo and O^{2-}) would change the configuration of the CFs and trigger a switching of the resistance state.

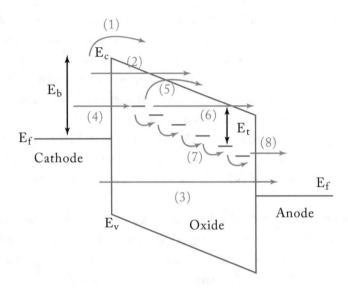

Figure 3.4: Schematic of the possible electron conduction paths through a metal/oxide/metal stack. (1) Schottky emission; (2) F-N tunneling; (3) direct tunneling; (4) tunneling from cathode to traps; (5) emission from trap to conduction band, which is essentially the Poole-Frenkel emission; (6) F-N-like tunneling from trap to conduction band; (7) trap-to-trap hopping or tunneling; and (8) tunneling from traps to anode. Adapted from [58].

To further characterize the electron trap/detrap processes in the defected oxide, the noise measurement is a powerful technique. A small constant stress voltage is applied on the RRAM cell, and its read-out current is sampled in the time-domain and then is transformed by the Fourier analysis into the frequency domain to obtain the $1/f^\alpha$ spectrum. In general, the current in HRS has a larger relative fluctuation. Figure 3.5 (a) shows the normalized noise power spectra density (S_I/I^2) in the frequency domain for different resistance states in the HfO$_x$-based RRAM [59]. It is seen that the higher the resistance state is, the larger the normalized noise power spectra density is. Also it is seen that for LRS the slope index α is close to 1, while for HRS there is a characteristic cutoff frequency f_0 above which α changes from 1 to 2. In general, the electron trap/detrap processes give a noise current on the top of the steady state current. The relaxation time τ (or the cutoff f_0) in the trap/detrap process is then determined by the transition time from the electrode to the traps [60]. Different traps with different distances from the electrode have different τ or f_0. Each trap gives a Lorentzian function with a specific cutoff frequency, as shown in Figure 3.5 (b). In LRS, CFs connect both electrodes, thus electrons can tunnel from the electrode to all the traps near the electrode with various relaxation times. Intuitively speaking, in LRS, the electrons have multiple choices when they tunnel from the electrode to the traps nearby, and the contribution from all these transitions will smooth the $1/f^2$ Lorentzian function and their envelope leads to $1/f$ behavior. In

HRS, the CF is partially ruptured and the shortest distance between the first trap and the electrode causes a minimum τ, thus $f_0 = 1/2\pi\tau_{(min)}$ corresponds to the cutoff frequency in the Lorentizian function. Therefore, the cutoff frequency becomes an indicator of the ruptured CF length. For a typical HRS range (500 kΩ-50 MΩ), the ruptured CF length is thus estimated to be 1.5 nm-2 nm. In the ultra-scaled RRAM devices, a single trap in the CF may dominate the conduction path. In this case, only $1/f^2$ behavior is shown, and in the time-domain, the random-telegraph-noise (RTN) can be observed [61, 62].

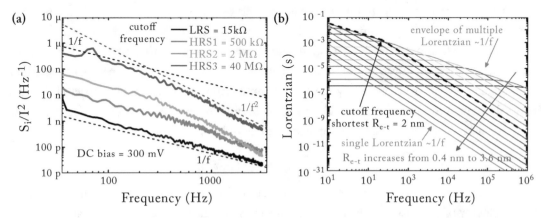

Figure 3.5: (a) Normalized noise power spectra density (S_i/I^2) in the frequency domain for different resistance states in the HfO$_x$-based RRAM. (b) Lorentzian function with different cutoff frequencies corresponding to different distances between the electrode and the trap. The contribution from tunneling from the electrode to multiple traps with various distances can smooth the envelope $1/f^2$ Lorentzian function to $1/f$. Adapted from [59].

3.3   NUMERICAL MODELING USING KINETIC MONTE-CARLO METHOD

To theoretically study the RRAM's intrinsic stochastic switching process, Kinetic Monte-Carlo (KMC) is a powerful method to model the atomistic processes such as Vo generation and O^{2-} migration and their recombination. As illustrated in Section 3.1, for the bipolar switching RRAM, during the forming/set process, O^{2-} are pulled out from lattice and Vo are generated and the CF forms connecting both electrodes. Then the current flows through the CF. During the reset process, the CF is partially ruptured by the recombination of Vo with the O^{2-} that migrate from the oxygen reservoir at the electrode/oxide interface; thus a tunneling gap is formed between the electrode and the residual CF. The electronic current conduction can typically be decoupled from the above ionic process in developing the RRAM model.

A 2D numerical KMC simulator based on the aforementioned principles has been developed [42] and applied for understanding the variability, current overshoot, and reliability degradation of the RRAM devices [63]. In the following, we will introduce an extended 3D numerical KMC simulator that can model the 3D filament evolution during the switching processes [64]. Figure 3.6 shows the simulation flow of the 3D KMC simulator: Starting with the Vo and O^{2-} distributions in the oxide layer, the electric field map is calculated by solving the 3D Poisson equation. Then the current is calculated with phonon-assisted TAT processes using the trap (i.e., Vo) locations. Considering the power dissipation from the energy released by the phonons, the 3D temperature profile is then calculated by solving the Fourier thermal transfer equation. Once the local temperature and electric field in the oxide layer are obtained, the Vo and O^{2-} distributions will be updated using a KMC method. Given a time step t, the probability of Vo and O^{2-} generation/migration/recombination events is calculated by the Boltzmann equation with an energy barrier (E_a) and a barrier-lowering term by enhanced field ($\gamma a_0 F$) at certain temperature (T) in the equation below. Note that here the electric field and temperature are both local to the specific sites of Vo and O^{2-}.

$$P(F, T, t) = \frac{t}{t_0} \exp\left(-\frac{E_a - \gamma a_0 F}{kT}\right) (0)$$

For different events, i.e., generation/migration/recombination, the barrier may be different. The probability of each Vo and O^{2-} for all possible events in the oxide layer will be calculated, and then the event selection will be performed using a random number by the KMC method. After the Vo and O^{2-} distributions are updated, it will move to the next time step until the program meets the stop conditions (i.e., the current reaches the compliance current, or the simulation period ends).

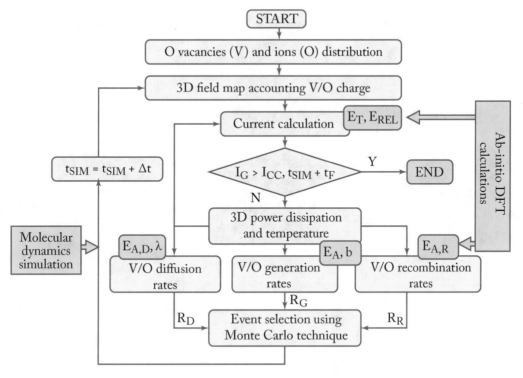

Figure 3.6: Simulation flow of the 3D Kinetic Monte-Carlo (KMC) simulator for modeling RRAM switching processes. Adapted from [64].

Figure 3.7 shows the simulated forming process for a TiN/Ti/HfO$_x$/TiN RRAM device. Initially, there are a few as-fabricated defects (i.e., Vo) in the cell. When the forming voltage is applied on the top electrode, the Vo and O^{2-} pairs are generated possibly along the grain boundary where initial defect density is higher, and then Vo concentrate and form the CF while O^{2-} migrate toward the top electrode and accumulate at the interface. Meanwhile, the temperature around the CF region can be raised up to more than 200°C. Figure 3.8 correlates the filament evolution with the simulated and experimental I-V curves of the devices. A full switching cycle from forming→ reset→ set is shown. During the reset, the O^{2-} migrate back from the interface and recombine with the Vo and rupture the CF with a formation of tunneling gap (the dashed region). In the subsequent set cycle, the CF is reformed but the shape of the CF varies from the previous cycle, which explains the stochastic nature of the resistive switching and the cycle-to-cycle variability in the switching parameters. It is worth pointing out that the location where the CF is ruptured is still under debate in literature. Some work [64] suggests that the CF is ruptured near the bottom electrode, while other work [63] suggests that the CF is ruptured near the top electrode. Further direct experimental observations are required to elucidate this uncertainty.

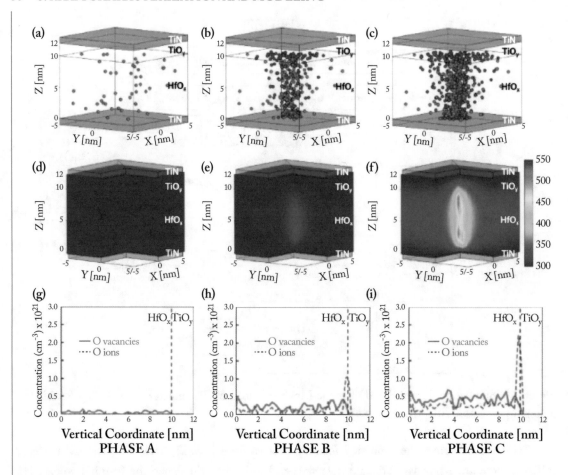

Figure 3.7: Simulated forming process for a TiN/Ti/HfO$_x$/TiN RRAM device using the 3D KMC simulator. (a)–(c) Vo (red) and O^{2-} (blue) distributions, (d)–(f) 3D temperature profile, (g)–(i) average Vo and O^{2-} concentration between the two electrodes during the three stages of the forming process. Adapted from [64].

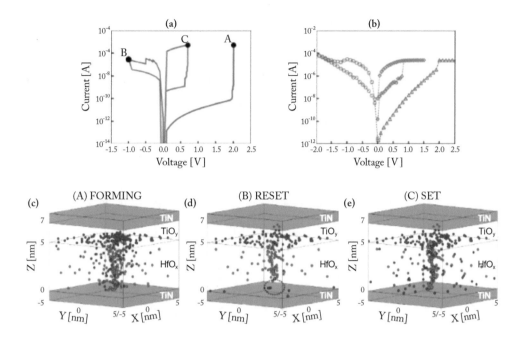

Figure 3.8: Simulated forming-reset-set process for a TiN/Ti/HfO$_x$/TiN RRAM device using the 3D KMC simulator. (a) Simulated (b) experimental I-V characteristics. Vo (red) and O^{2-} (blue) distributions at the end of (c) forming, (d) reset, and (e) set operations. Adapted from [64].

Besides the KMC method to model the atomistic processes for discrete Vo and O^{2-}, the other numerical approach is to consider the concentration or density of Vo or O^{2-}, and solve the drift-diffusion equation for the Vo profile under the electric field and concentration gradient [65, 66]. Then the Vo profile determines the conductivity of a resistor network. This approach is numerically simpler than the KMC method but it is based on the concept of the "continuous" Vo profile, which may become problematic when the formation and rupture of the CF occurs within a distance of one or two nanometers.

3.4   COMPACT MODELING FOR SPICE SIMULATION

In order to facilitate the circuit-level design, a compact model of RRAM that can be run in the SPICE simulation engine is very useful. In literature, there are a several existing RRAM compact models [67, 68, 69, 70] that used a simplified physical picture of the CF formation and rupture. Here we present a representative model that has been calibrated with IMEC's HfO$_x$-based RRAM, which can be publicly downloaded at [71]. Figure 3.9 shows the RRAM device structure in the compact model. In this model, a single dominant CF in one dimension is considered. The primary internal variable used in this model is the gap distance (g), which is defined as the average distance between the top electrode (TE) and the tip of the CF. g can determine the RRAM resistance through the electron tunneling conduction mechanism, where the resistance exponentially increases with g. Besides, the resistance also has the nonlinear relationship with the applied voltage (V). Generally, the RRAM resistance shows linear dependence at small V (typically <0.5 V) and exponential dependence at large V, which can be modeled by a hyper-sinusoidal function. Thus, the I-V relationship of the RRAM model is expressed as

$$I = I_0 \exp\left(-\frac{g}{g_0}\right) \sinh\left(\frac{V}{V_0}\right) \tag{1}$$

where I_0, g_0 and V_0 are fitting parameters to a specific set of the RRAM I-V data.

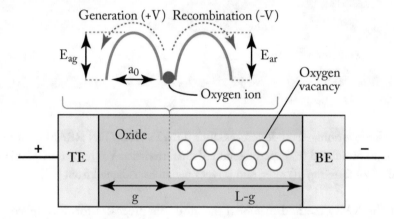

Figure 3.9: Schematic of RRAM device structure and the generation/recombination process of oxygen ion at the conductive filament tip.

The RRAM switching is essentially a dynamic process, even with a fixed V, the current I is time-dependent as g evolves. By applying a positive (negative) voltage on the RRAM device, the CF will grow (dissolute) due to the generation (recombination) of Vo and O^{2-} at the tip of the CF, which can be expressed with the following equations

$$\frac{dg}{dt} = -v_0 \left[\exp\left(-\frac{qE_{ag}}{kT}\right) \exp\left(\frac{\gamma a_0}{L}\frac{qV}{kT}\right) - \exp\left(-\frac{qE_{ar}}{kT}\right) \exp\left(-\frac{\gamma a_0}{L}\frac{qV}{kT}\right)\right] \tag{2}$$

$$\gamma = \gamma_0 - \beta \left(\frac{g}{g_1}\right)^3 \tag{3}$$

$$g(t+dt) = g(t) + dg \tag{4}$$

where dg/dt is the gap growth/dissolution velocity. g has a lower and an upper limit on the average gap size g_{min} and g_{max}, respectively. g_{min} represents the tip of CF nearly in contact with the TE during the set operation. In this case, the resistance at the TE interface layer may become the dominant player. g_{max} stands for the residual CF which cannot be removed anymore during the reset operation. In Equation (2), dg/dt can be calculated by the net difference between the Vo generation and recombination rates. E_{ag} (E_{ar}) is the activation energy for O^{2-} to migrate from one potential well to another in the generation (recombination) process. If E_{ag} is not equal to E_{ar}, the filament will still be gradually changing even under zero voltage bias, which accounts for the ion self-diffusion process. Normally E_{ag} should be larger than E_{ar}, in order to capture the dominant retention failure mode (LRS resistance shift toward HRS) as reported in [45]. L is the oxide thickness, and a_0 is the atomic hopping distance. $a_0(qV/L)$ can be viewed as the energy barrier raising/lowering to the neighboring oxygen vacancy site. γ is the g-dependent local field enhancement factor, which is calculated in Equation (3). It considers the polarizability of high-k dielectrics and non-uniform potential distribution in the device structure [72]. The form of (3) is obtained from empirical fitting of the HfO_x-based RRAM devices under pulse train measurement for a gradual RESET process [73], which introduces a positive (negative) feedback on the filament growth (dissolution) for abrupt set and gradual reset behaviors in typical bipolar RRAM devices. v_0, γ_0, β, and g_1 are fitting parameters. T is the local temperature of the CF, and the T evolution can be expressed by a simplified heat conduction process

$$\frac{dT}{dt} + \frac{T-T_0}{\tau_{th}} = \frac{|V \times I|}{C_{th}} \tag{5}$$

$$T(t+dt) = T(t)+dT \tag{6}$$

In Equation (5), T_0 is the ambient temperature, and τ_{th} and C_{th} are the effective thermal time constant and thermal capacitance, respectively. The above equations are implemented in the Verilog-A language for the compatibility with the SPICE simulator. The model parameters are calibrated from IMEC's $TiN/Hf/HfO_x/TiN$ based RRAM devices [43, 45]. Figure 3.10 (a) shows the simulated and experimental quasi-DC I-V curve, and Figure 3.10 (b) shows the simulated and experimental retention degradation at high temperature. To simulate the pulse programming conditions in the one-transistor and one-resistor (1T1R) configuration, the transistor is implemented using the PTM model [74] at 130 nm technology, to match the channel length that was used in IMEC's test structure. The summary of the pulse programming conditions fitting is shown in Table 3.1. The discrepancy in the gate voltage of the series transistor (V_{WL}) is due to a mismatch of the drivability of the PTM transistors with the specially processed IMEC transistors.

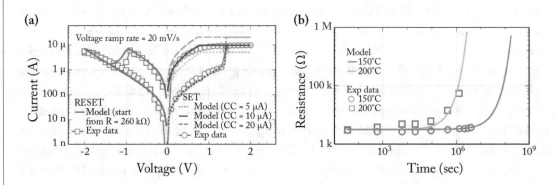

Figure 3.10: (a) Quasi-DC I-V fitting with IMEC HfO$_x$-based RRAM. Different compliance currents (CC) are applied in the set process to demonstrate the modulation of the LRS resistance by the compliance current. (b) LRS retention time fitting with IMEC HfO$_x$-based RRAM. Experimental data from [45].

Table 3.1: Pulse programming fitting in 1T1R configuration with compact model [71]

	IMEC	Model
SET (R= 100 kΩ to R= 10 kΩ)	V_{WL}= 1 V V_{BL}= 1.8 V SET time<= 10 ns	V_{WL}= 1.4 V V_{BL}= 1.8 V SET time≈ 10 ns
RESET (R= 10 kΩ to R= 100 kΩ)	V_{WL}= 3 V ~ 3.5 V V_{SL}= 1.6 V ~ 2.5 V RESET time<= 10 ns	V_{WL}= 3 V V_{SL}= 1.8 V RESET time≈ 7.5 ns

CHAPTER 4

RRAM Array Architecture

4.1 1T1R ARRAY

One of the common RRAM array architectures is the one-transistor and one-resistor (1T1R) array. In this design each RRAM cell is in series with a cell selection transistor, as shown in Figure 4.1. The addition of a selection transistor is able to isolate the selected cell from other unselected cells. The word line (WL) controls the gate of the transistor, thus tuning the WL voltage can control the compliance current that is delivered to the RRAM cell. The RRAM cell's top electrode connects to the bit line (BL) while its bottom electrode connects to the contact via of the drain of the transistor. The source line (SL) connects to the source of the transistor. The typical cell area of 1T1R array is 12 F^2 (F is the lithography feature size) if the gate width/length (W/L) of the transistor is 1. The minimum cell area can be reduced to 6 F^2 if the aggressive borderless DRAM design rule with sharing BL and SL is applied. The cell area will be increased if W/L of transistor is larger than 1 when a minimum sized transistor cannot provide sufficient programming current. Because of a relatively large cell area but a good isolation between cells that minimizes the cross-talk problem, 1T1R array is preferred for embedded applications where the density is not the pursuit but the performance and the reliability are of the priority.

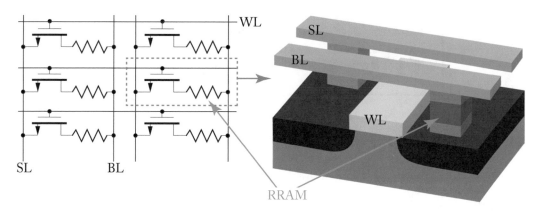

Figure 4.1: Schematic of the 1T1R array architecture.

ITRI has reported a 4 Mb 1T1R HfO_x-based RRAM prototype chip [35], as shown in Figure 4.2. The fabrication was done in 180 nm CMOS process. Single-level-cell (SLC) operation with 7.2 ns read/write random access has been presented, and multi-level-cell (MLC) 2 bit/cell operation with 160 ns write-verify scheme has been demonstrated.

Process	CMOS: 0.18 μm 1P4M RRAM: 0.64 μm x0.48 μm
Memory Capacity	4 Mb (32 ×128Kb sub-blocks)
Chip Size	11310 μm × 16595 μm (with test-mode circuits)
Device	HV path: 3.3 V device Cell array: 3.3 V device Peripheral: 1.8 V device
VDD	HV path: 3.3 V Core: 1.8 V
Read-Write Access Time (SLC-mode)	Random access: 7.2 ns Burst-mode: 3.6 ns

Figure 4.2: ITRI's 4 Mb 1T1R HfO$_x$-based RRAM prototype chip. Adapted from [35].

Figure 4.3 shows the typical write/read scheme for the 1T1R array. For the set operation, WL voltage is applied to turn on the transistor of the selected cell, and a write voltage is applied to the BL of the selected cell while SL is grounded; for the reset operation, WL voltage is applied to turn on the selection transistor of the selected cell, and a write voltage is applied to the SL of the selected cell while BL is grounded to reverse the current, as the typical RRAM operation needs bipolar switching. For unselected rows and columns, the WL, BL, and SL are all grounded. To read out the data from the 1T1R array, WL voltage is applied to turn on the selection transistor of the selected cell, and a read voltage is applied to the BL while SL is grounded. The sense amplifier thus can sense the difference in the read-out current for HRS and for LRS through the BL with a reference. Because the transistors are off for the unselected cells, there is no cross-talk or interference issues in the 1T1R array, and each cell can be independently and randomly accessed. Multiple-bits can be written (or read) in parallel into (or from) the same row by activating multiple columns.

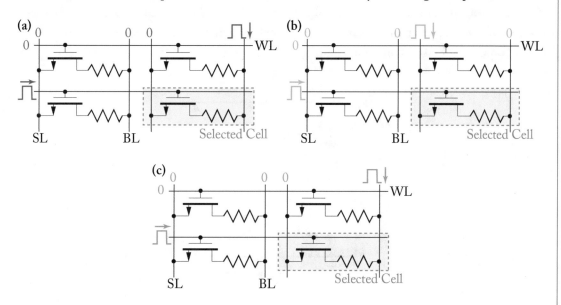

Figure 4.3: Write/read scheme for 1T1R array (a) set scheme, (b) reset scheme, and (c) read scheme.

Conventional designs may use different WL voltages for set and reset, and typically the reset WL voltage is larger than the set WL voltage, because part of the reset WL voltage is dropped on the RRAM cell; thus a larger WL voltage is needed to turn on the transistor. Due to different WL voltages, the set and reset operations cannot be performed simultaneously on the same selected row. Therefore, a two-step write process is needed if a string of multiple bits is written into the 1T1R array: first set the "1" bits and then reset the "0" bits. In order to speed up the parallel write operation for multiple bits, the same WL for both set and reset can be used by designing appropriate voltage settings [75].

The 1T1R array may face scaling challenges if the RRAM's programming current or programming voltage cannot scale accordingly. Although the RRAM cell itself has excellent scalability down to sub-10 nm regime as discussed in Section 2.1, RRAM's programming current typically does not scale with the device area due to the filamentary conduction mechanism. Figure 4.4 shows the silicon CMOS low-power logic transistor's drive current with the scaling from 130 nm down to 10 nm for different W/L simulated with the PTM model [74]. If the RRAM's programming current keeps today's representative value ~50 µA, W/L=2 is needed for 65 nm beyond and W/L=3 is needed for 22 nm beyond. If the RRAM's programming current keeps today's optimized value ~20 µA, W/L larger=2 is still needed for 32 nm beyond. Although the transistor's gate voltage can be boosted to increase the drive current, the large gate voltage inevitably introduces the gate dielectric reliability issues. Moreover, if the RRAM's programming voltage keeps today's typical value ~2 V under pulse mode, then the reversed p-n junction of the transistor's bulk to drain may face large reverse leakage current and breakdown issue as well. Therefore, reducing RRAM's programming current down to sub-10 µA and programming voltage down to sub-1 V by device engineering is of great importance for the scaling of the 1T1R array.

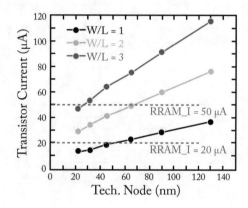

Figure 4.4: Silicon CMOS low-power logic transistor's drive current with the scaling from 130 nm down to 10 nm for different W/L simulated with the PTM model [74].

4.2   CROSS-POINT ARRAY

One of the common RRAM array architectures is the cross-point (or crossbar) array, which consists of rows and columns perpendicular to each other with RRAM cells sandwiched in between, as shown in Figure 4.5. The cross-point array supports a 4 F^2 cell area, thus it can achieve higher integration density than the 1T1R array. For the standalone and large capacity NVM, the cross-point array is more attractive. The scalability of the cross-point array is not limited by the drivability of the cell selection transistors as the case in the 1T1R array. The programming current is provided by the driver transistors at the edge of the cross-point array, of which the W/L can be increased at advanced technology nodes with affordable area overhead. The driver should provide sufficient current for the programming current of the selected cell in addition to the sneak path current of the unselected cells. A selector device is typically added in series with the RRAM cell at each cross-point to enable a large-scale cross-point array by cutting off the sneak path current of the unselected cells, and the detailed discussions on the sneak path problem and the selector device design will be presented in the next Section 4.3. In the following two prototype chip demonstrations, the selectors are employed.

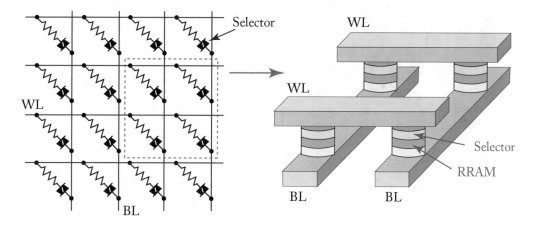

Figure 4.5: Schematic of the cross-point array. Selector is added in series with the RRAM cell at each cross-point.

Panasonic has reported a 8 Mb cross-point TaO_x-based RRAM prototype chip [36], as shown in Figure 4.6. The fabrication was done in 180 nm CMOS process. 443 MB/s write throughput (64-bit parallel write per 17.2 ns cycle) and 25 ns read access have been demonstrated. Sandisk/Toshiba has reported a 32 Gb cross-point MeO_x[6]-based RRAM prototype chip [37], as shown in Figure 4.7. Both Panasonic and Sandisk/Toshiba's design adopted a 2-layer stacked cross-point array architecture by sharing the BL to increase the integration density, as shown in Figure 4.8 (a). Figure 4.8 (b) shows the cross-section TEM image of Sandisk/Toshiba's prototype chip. It can be seen that the RRAM cell size is very small (24 nm) by BEOL lithography, however its peripheral circuits underneath the cross-point array still uses an older technology node (possibly 130 nm or 180 nm). Therefore, in this design one sub-array can only have one sense-amplifier, and it requires an external sensing scheme that borrows sense-amplifiers from other sub-arrays. As a result, the read latency is limited to be 40 μs. An array macro modeling work using the NVSim simulator [76] suggests that by scaling the peripheral circuits technology node with the RRAM cell together, a 2-layer cross-point array at the 10 nm node is projected to achieve ultra-high density ~3.43 Gb/mm^2, and can enable fast write bandwidth ~ 300 MB/s and read bandwidth ~1 GB/s [77]. As a reference, the planar 64 Gb NAND FLASH chip with MLC at 1x nm node[7] has a density ~ 0.585 Gb/mm^2, and write bandwidth ~ 33 MB/s [78].

---

[6] The metal oxide materials are undisclosed by Sandisk/Toshiba, thus MeO_x is used here.

[7] In planar NAND FLASH development, 1x nm node is typically ~18 or 19 nm, and 1y nm node is typically ~15 or 16 nm. This definition varies among companies.

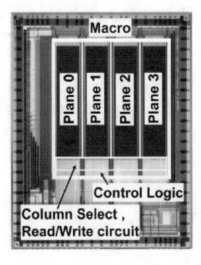

Process	0.18 µm 1Poly 4 Metal
Memory Cell	2-layered cross-point 0.38 µm × 0.38 µm Ir/Ta$_2$O$_5$/TaO$_x$/TaN/SiN$_x$/TaN Resistive memory element Diode
Capacity	8Mbit (64 Kpage x 64 bits x 2 layers)
Macro	3.0 mm x 3.5 mm
I/O	16 bits (internal 64 bits)
ECC	7 Parity bits for 64 normal bits
Write	443 MB/s
Read	25 ns

Figure 4.6: Panasonic's 8 Mb cross-point TaO$_x$-based RRAM prototype chip. Adapted from [36].

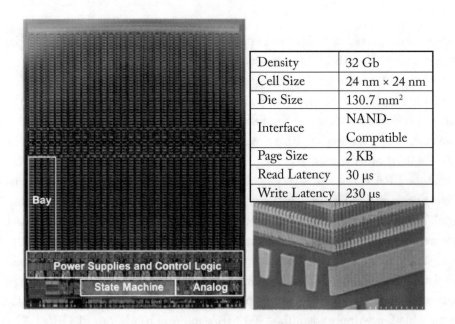

Density	32 Gb
Cell Size	24 nm × 24 nm
Die Size	130.7 mm^2
Interface	NAND-Compatible
Page Size	2 KB
Read Latency	30 µs
Write Latency	230 µs

Figure 4.7: Sandisk/Toshiba's 32 Gb cross-point MeO$_x$ [8]-based RRAM prototype chip. Adapted from [37].

---

[8]   The metal oxide materials are undisclosed by Sandisk/Toshiba, thus MeO$_x$ is used here.

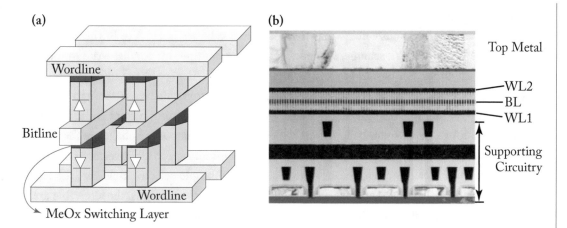

Figure 4.8: (a) Sandisk/Toshiba's 2-layer cross-point array architecture by sharing the BL. (b) The cross-section TEM of their 32 Gb RRAM prototype chip. Adapted from [37].

The write/read schemes of the cross-point array are discussed as follows. Because of no cell selection transistors, cross-talk or interference exists between cells in the cross-point array. To successfully program the RRAM cells, two typical write schemes (V/2 scheme and V/3 scheme) can be applied. Figure 4.9 (a) shows the voltage bias conditions for the V/2 scheme. In the V/2 scheme, for the set operation, the selected cell's WL and BL are biased at the write voltage V_W and ground, respectively. For the reset operation, the bias conditions on WL and BL are reversed for the bipolar switching. In both set and reset operations, all the unselected WLs and BLs are biased at $V_W/2$. Therefore, only the selected cell sees a full V_W, while the half-selected cells along the selected WL or BL see a half V_W, and all the other unselected cells in the array see zero voltage (in reality, due to the IR drop along interconnect, the voltage is not perfectly zero though). Here the assumption is that half Vw should not disturb the RRAM's resistance. Figure 4.9 (b) shows the voltage bias conditions for the V/3 scheme. In the V/3 scheme, for the set operation, the selected cell's WL and BL are biased at the write voltage V_W and ground, respectively. For the reset operation, the bias conditions on WL and BL are reversed for the bipolar switching. The unselected WLs and BLs are biased at 1/3 V_W and 2/3 V_W for the set operation, respectively. The unselected WLs and BLs are biased at 2/3 V_W and 1/3 V_W for the reset operation, respectively. In this way, the selected cell sees V_W, while all other unselected cells in the array only see 1/3 V_W. Here the assumption relaxes so that 1/3 V_W should not disturb the RRAM's resistance.

The pros and cons of these two write schemes can be summarized as follows: the V/2 scheme typically has less power or energy consumption than the V/3 scheme. This is because the unselected cells (not along the selected WL and BL) in the V/2 scheme see zero voltage ideally, while all the unselected cells in the V/3 scheme see 1/3 V_W, thus consuming static power during the write period. On the other hand, the V/3 scheme has better immunity to the write disturbance than the V/2 scheme, as the maximum voltage that the unselected cells see is 1/3 V_W in the V/3 scheme, while it is 1/2 V_W in the V/2 scheme. It is possible to have multiple-bit parallel write in the cross-point array with either V/2 or V/3 scheme by biasing multiple BLs (or WLs) to be ground in the set (or reset) operation. However, unlike the programming current delivered through BL or SL independently for multiple columns in the 1T1R array, the programming current is shared through the same WL for multiple columns in the cross-point array. This imposes a challenge for the W/L for driver transistors at the edge of the array. One driver (typically a CMOS inverter or an inverter chain) needs to deliver the programming current for the multiple selected cells and the leakage current for the other half-selected cells at the same row. In practical designs, the allowable area or the drivability of the driver may limit the number of the cells that can be written in parallel.

Figure 4.10 shows the read scheme for the cross-point array. All the columns are biased at the read voltage V_R, while the selected row is biased at ground and the unselected rows are biased at V_R. Therefore, only the cells of the selected row see a read voltage and all the other unselected cells see zero voltage (in reality, due to the IR drop along interconnect, the voltage is not perfectly zero though). The entire selected row can be read-out in parallel if each column can have one sense amplifier. In practical designs, the sense amplifier is large in area, thus multiple columns have to share one sense amplifier.

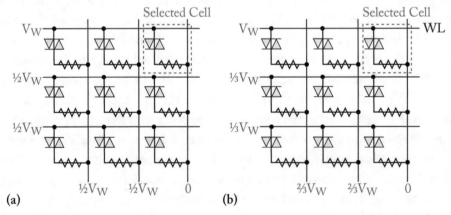

Figure 4.9: (a) V/2 write scheme and (b) V/3 write scheme for the cross-point array.

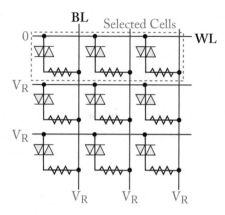

Figure 4.10: Read scheme (for an entire row) for the cross-point array.

4.3   SELECTOR DEVICE

The cross-point array suffers two well-known design challenges: (1) IR drop problem along the interconnect wires and (2) sneak path problem through the unselected cells, as shown in Figure 4.11. The IR drop problem becomes significant when the WL and BL wire width scales to sub-50 nm regime where the interconnect resistivity drastically increases due to the electron surface scattering. For example, at 20 nm node, the copper interconnect resistance between two neighboring cells is 2.93 Ω; thus the IR drop along the wire for a large array (e.g., 1024 × 1024 array) is no longer negligible, as the farthest cell from the driver sees an interconnect resistance ~3 kΩ. If the RRAM cell's LRS resistance is comparable to this interconnect resistance (typically tens of kΩ), a portion of the write voltage will drop on the wire instead of the RRAM cell. To guarantee a successful write operation, the write voltage provided from the driver has to be boosted over the actual switching voltage of the RRAM cell to compensate the IR drop. However, the write voltage cannot be boosted too much because 1/2 V_W (in the V/2 scheme) should not disturb the RRAM resistance for the cells close to the driver. The sneak path problem is associated with the IR drop problem. Taking the V/2 scheme as an example, the half-selected cells along the selected WL and BL conduct the leakage current and form the sneak paths during the write operation. The sneak paths contribute current to the IR drop and further degrade the write margin. The sneak path problem also degrades the read margin during the read operation: if the cell to be read out happens to be in HRS with surrounding unselected cells in LRS, the sneak path current can flow through the surrounding cells in the LRS and sink to the read-out path, therefore a current higher than the actual HRS current is read out which decreases the sense margin between HRS and LRS. The sneak paths exacerbate the degradation of read margin when the rows or the columns are floating, thus it is preferred to fix the WL and the BL voltages as shown in the aforementioned read scheme in Figure 4.10. If there is no interconnect resistance, the sneak paths actually do not exist in the read operation in the aforementioned read scheme in Figure 4.10. However, in reality, the sneak paths are inevitable because the interconnect resistance makes the voltage across the unselected cells non-zero. Further discussions about the IR drop problem and the sneak path problem of the cross-point array architecture can be referred to [79, 80, 81]. The conclusions from these works indicate that increasing the LRS resistance (or equivalently reducing the write current) and increasing the I-V nonlinearity of the RRAM cell (with the help of the selector) are useful to minimize the IR drop and sneak paths, thereby magnifying the write/read margin. If the LRS resistance is much larger than the interconnect resistance, most of the write voltage drops on the RRAM cell. However, a trade-off exists between the write margin and the read margin. A larger LRS resistance also indicates a smaller read-out current, which requires a longer time for sensing. As a reference, state-of-the-art current-mode sense amplifier can sense sub-100 nA read-out current within 26 ns [82]. Therefore, the upper-limit of the LRS resistance is limited by the read-out current level.

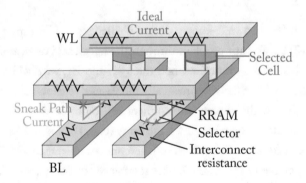

Figure 4.11: Schematic of the sneak path problem and IR drop problem in the cross-point array.

In order to further suppress the sneak paths, a selector with strong I-V nonlinearity is desired to be added to the RRAM cell [83]. The selector can be a diode for unipolar switching RRAM as the one-diode and one-resistor (1D1R) architecture, or a bi-directional selector (with a strong nonlinear I-V characteristics in both polarities) for bipolar switching RRAM as the one-selector and one-resistor (1S1R) architecture. The selector devices effectively suppress the leakage current at reversed bias for the unipolar switching RRAM or at low bias (e.g., $V_W/2$ for the V/2 scheme) for the bipolar RRAM, thus the interference between neighboring cells is prevented. We will now survey the two-terminal selector devices reported in literature.

For unipolar switching, a p-n diode is the most common device for the cell selector. Although a high performance p-n diode is easily fabricated with current epitaxial silicon technology for the planar device structure, it is not feasible to implement epitaxial silicon-based p-n diode with the RRAM array at the back-end-of-line (BEOL) because it is difficult to grow epitaxial silicon on a metal layer, and high processing temperature is required. On the other hand, amorphous silicon allows for lower processing temperature. But it does not meet the requirement for the current density for the RRAM programming. Therefore, new device structures need to be explored for the cell selector, which should both allow for low processing temperature and also provide high current drivability. Compared to silicon p-n diode, oxide p-n diode is more attractive because it offers better flexibility in processing technology and it can be fabricated during BEOL processing even at room temperature. If the oxide material is oxygen deficient with a sufficient amount of oxygen vacancies, it is n-type; while if the oxide material is metal deficient with a sufficient amount of metal vacancies, it is p-type. Thus a combination of p-type oxide and n-type oxide essentially forms a p-n diode. Several kinds of oxide p-n diodes [84], such as p-NiO/n-TiO$_2$, p-NiO/n-ZnO, p-NiO/n-InZnO, p-CuO/n-InZnO, have been demonstrated and stacked with Pt/NiO/Pt RRAM in series, among which p-CuO/n-InZnO is regarded as the best candidate in terms of current drivability. Besides the p-n oxide diode, through oxide/electrode interface engineering, rectifying I-V for unipolar switching can also be achieved. For example, Schottky diode Pt/TiO$_2$/Ti/Pt stack has been integrated with Pt/TiO$_2$/Pt unipolar RRAM [85].

For bipolar switching, bi-directional nonlinearity is required. Oxide/electrode interface engineering or oxide/oxide bandgap engineering with the tunneling current mechanism can be leveraged as the tunneling current generally increases exponentially with the applied voltage. For example, Ni/TiO$_2$/Ni bi-directional selector has been integrated with Ni/HfO$_x$/Pt bipolar RRAM [86], and Pt/TaO$_x$/TiO$_2$/TaO$_x$/Pt bi-directional selector has been integrated with Cu/HfO$_x$/Pt bipolar RRAM [87]. In addition, Cu ion motion in the Cu-containing Mixed Ionic Electronic Conduction (MIEC) materials also shows a good bi-directional nonlinearity for bipolar switching RRAM, as demonstrated by IBM's series work [88, 89, 90]. The aforementioned selectors rely on an exponential slope in the I-V curve to turn on the selector accompanied with an increase of the current by several orders of magnitude. Ideally, an abrupt turn-on behavior with minimal transition voltage is preferred, which is referred to as the threshold switching. This can be achieved in the metal-insulator-transition (MIT) materials such as VO$_2$, NbO$_2$. Unlike RRAM devices, the threshold switching behavior is not bistable and can occur at both voltage polarities. The threshold selector device will be turned on above a threshold voltage and will be turned off below a hold voltage. For example, Pt/VO$_2$/Pt selector has been integrated with NiO unipolar RRAM [91] and ZrO$_x$/HfO$_x$ bipolar RRAM [92]. However, the VO$_2$ has a transition temperature around 67°C beyond which the threshold switching behavior disappears [93], which is a major drawback for practical applications. Alternatively, NbO$_2$ has a transition temperature around 800°C, thus it is more attractive due to its thermal stability. TiN/NbO$_2$/W selector has been integrated with TaO$_x$ bipolar RRAM [94]. The drawback of MIT-based threshold selector is a relatively small on/off rectify ratio. Recently, a new kind of threshold selector named Field Assisted Superlinear Threshold (FAST)[9] selector has been reported [95], which shows outstanding on/off rectify ratio ($>10^7$), small turn-on slope (<5 mV/dec), and high current drivability (>5 MA/cm^2), and the threshold voltage is claimed to be adjustable from ~0.3 V to ~1.3 V to match various RRAM characteristics. Figure 4.12 (a) and (b) show the I-V characteristics of the RRAM cell and the FAST selector, respectively, and Figure 4.12 (c) shows the I-V characteristics of the stacked FAST selector with the RRAM cell (1S1R). It can been seen that if the write voltage is designed to be 2 V for the 1S1R cell, then $V_W/2 = 1$ V will turn off the selector and substantially suppress the sneak paths. The read voltage can be designed to be 1.5V to make the read-out current's on/off ratio $>100\times$.

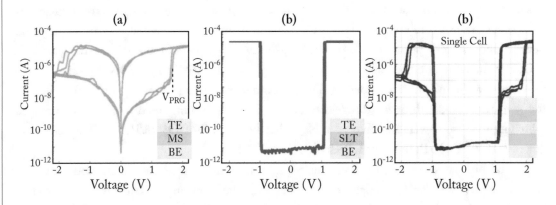

Figure 4.12: (a) I-V characteristics of the RRAM cell and (b) I-V characteristics of the FAST selector. (c) I-V characteristics of the stacked FAST selector with the RRAM cell (1S1R). Adapted from [95].

Table 4.1 compares several selector devices reported in literature in aspects of voltage range, current drivability, rectify or nonlinearity ratio. Although substantial progress has been made in the past few years, the development of selector devices is still a key challenge for implementing cross-point memory architecture today. Some of the selectors use Pt in the electrode, which is not CMOS fabrication process-friendly. More importantly, the selector device characteristics must match the RRAM device characteristics. Adding selector device in series with the RRAM cell inevitably increases the programming voltage as part of the voltage is used to turn on the selector device. For the bi-directional selector with exponential I-V, the read sense margin generally degrades because the on-state or LRS current for read-out is also suppressed. For the threshold switching selector with abrupt I-V, the read voltage has to been boosted above the threshold voltage of the selector; thus it runs a risk of disturbing the RRAM resistance in the read operation. Ideally, the RRAM cell itself is preferred to have built-in I-V nonlinearity, thus eliminating the necessity of the external selector device.

---

[9] The FAST selector's materials are undisclosed by the developer Crossbar, Inc.

Table 4.1: Representatives of selector devices for RRAM reported in literature

Type	Stack	Voltage Range	Current Drivability	Rectify Ratio
p-n diode	p-CuO/n-InZnO [84]	-2V~+2V	3×10^4 A/cm^2	3×10^4
Schottky diode	Pt/TiO$_2$/Ti/Pt [85]	-2V~+2V	3×10^5 A/cm^2	2.4×10^6
Bi-directional selector	Ni/TiO$_2$/Ni [86]	-4V~+4V	10^5 A/cm^2	10^3
Bi-directional selector	Pt/TaO$_x$/TiO$_2$/TaO$_x$/Pt [87]	-2.5V~+2.5V	3.2×10^7 A/cm^2	10^4
Bi-directional selector	MIEC [90]	-1.6V~+1.6V	50×10^6 A/cm^2	10^4
Threshold selector	Pt/VO$_2$/Pt [92]	-0.5V~+0.5V	6×10^6 A/cm^2	50
Threshold selector	TiN/NbO$_2$/W [94]	-1V~+1V	10×10^6 A/cm^2	50
Threshold selector	FAST [95]	-1V~+1V	5×10^6 A/cm^2	10^7

The voltage range is the max voltage where the current density is measured, for the threshold selector, the voltage range is the threshold voltage; for diode, the rectify ratio is defined as the forward/reverse current at the max voltage; for bi-directional and threshold selector, the rectify ratio is defined as the current at max voltage over half of the max voltage.

4.4   PERIPHERAL CIRCUITS DESIGN

A typical RRAM sub-array (or block) has the following peripheral circuits in addition to the RRAM array core: row decoders and WL drivers, BL multiplexer, sense amplifiers, sense amplifiers multiplexers, and output or write drivers, as shown in Figure 4.13 (a). There are two approaches for integrating RRAM cells on top of the CMOS circuits. The first approach is to fabricate the RRAM cells following the front-end-of-line (FEOL) process (close to the transistor fabrication at lower-level interconnect). For example, the RRAM cells can be deposited at the contact via between the drain and Metal 1, and this approach is typically employed in the 1T1R array architecture. The second approach is to fabricate the RRAM cells via the back-end-of-line (BEOL) process at top-level interconnect (decoupled from the transistor fabrication). For example, the RRAM cells can be deposited at the contact via between Metal 4 and Metal 5, and this approach is typically employed in the cross-point array architecture. One of the advantages of the BEOL integration is that the peripheral circuits can be hidden underneath the cross-point array to save the area as demonstrated in Panasonic's 8 Mb prototype chip [36], as shown in Figure 4.13 (b).

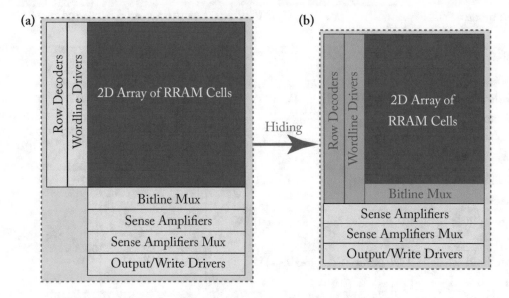

Figure 4.13: Schematic of a typical RRAM sub-array (or block) with the peripheral circuits. For cross-point array, part of the peripheral circuits can be hidden underneath of the array core as the RRAM cells can be fabricated at top-level interconnect layers through BEOL processing.

Besides the yield and variability issues of the RRAM technology, another challenge for the peripheral circuit design and its integration with RRAM is the incompatibility of the RRAM's programming voltage and the transistor's supply voltage (VDD). This problem is exacerbated when the RRAM cells are embedded into the logic-process, when the VDD of today's transistors below 45 nm reduces to sub-1 V regime while most of the RRAM devices still retain programming voltage in the range of 1~3 V in the pulse mode. Further reduction of the RRAM programming voltage close to the nominal VDD of the I/O-devices or core-devices is necessary.

One of the key components of the peripheral circuits is the sense amplifier (S/A) for reading-out the RRAM memory states. Here we briefly introduce the designs of the sense amplifier following the discussions in [96]. The sense amplifiers can be generally categorized into two types: voltage-mode sensing and current-mode sensing. Figure 4.14 shows the voltage-mode sensing scheme and the schematic of the corresponding waveforms. The operation of the voltage-mode sensing is divided into three phases: BL precharge, BL voltage development, and voltage comparison. In the BL precharge phase, a precharge transistor is turned on to increase the BL voltage from

0 V to a precharge voltage. In the BL developing phase, BL voltage tends to decay with different slopes corresponding to different memory states of the selected RRAM cell. When an HRS cell is read, the read-out current is relatively small, thus the BL voltage is maintained near the precharge voltage. When an LRS cell is read, a larger read-out current causes the BL voltage to discharge faster, which generates a BL voltage swing larger than that of the HRS cell. In the voltage comparison phase, once the BL voltage swing is developed with sufficient sense margin, the S/A_EN will activate a voltage-mode sense amplifier (VSA) (e.g., a differential amplifier with the latch load), which compares the BL voltage with a reference voltage and generates a digital output at DOUT.

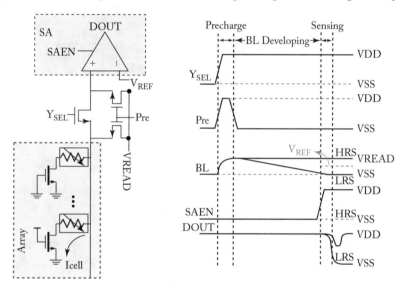

Figure 4.14: The voltage-mode sensing scheme and the schematic of the corresponding waveforms. Adapted from [96].

Figure 4.15 shows the current-mode sensing scheme and the schematic of the corresponding waveforms. In the voltage-mode sensing, the BL voltage decays over time; however, in the current-mode sensing scheme, the BL voltage is maintained a constant value (BL clamping voltage). The operation of current-mode sensing is similarly divided into three phases: BL precharge, current development, and current comparison. In the BL precharge phase, a precharge transistor is turned on to increase the BL voltage from 0 V to the BL clamping voltage. In the cell current development phase, the cell read-out current is different under a constant BL clamping voltage corresponding to different memory states in the selected RRAM cell. The read-out current of an LRS cell is larger than that of an HRS cell. Finally, in the current comparison phase, a current-mode sense amplifier

(CSA) compares the read-out current of the selected cell using a reference current and generates a digital output at DOUT.

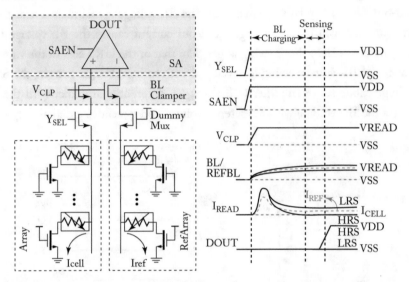

Figure 4.15: The current-mode sensing scheme and the schematic of the corresponding waveforms. Adapted from [96].

In practical designs, the choice between the voltage-mode sensing and the current-mode sensing depends on the array size and the RRAM cell characteristics. Figure 4.16 shows a comparison of the sensing speed for voltage-mode sensing and current-mode sensing with different BL lengths. For an array with a long BL length or a higher LRS resistance, current sensing provides faster access. Voltage sensing is preferred when the BL length is short or the LRS resistance is small.

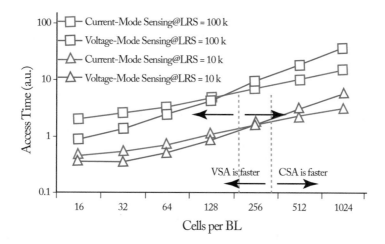

Figure 4.16: A comparison of the sensing speed for voltage-mode sensing and current-mode sensing with different BL lengths. Adapted from [96].

4.5   3D INTEGRATION

The primary target of RRAM is to replace the NAND FLASH technology for standalone large-capacity storage, as NAND FLASH is facing scaling limitations beyond 10 nm technology node. State-of-the-art 2D NAND FLASH has been scaled down to around 15 nm in 2015, while the 3D stackable NAND FLASH is emerging [97, 98] and 24-layer up to 32-layer 128 Gb 3D NAND FLASH chip featuring MLC has been demonstrated [99, 100] and 3D NAND FLASH based solid-state drive (SSD) has been commercialized. Although at the single device level, RRAM outperforms NAND FLASH in many aspects such as much faster programming speed, smaller programming voltage, better reliability, etc., the key challenge for RRAM to compete with NAND FLASH is the integration density, or more importantly, the cost per bit. To achieve similar device density as the 3D NAND FLASH, a technology path toward the 3D stackable RRAM is required.

There are two 3D integration approaches for RRAM technology [101]: one is based on stacking the conventional horizontal RRAM array layer by layer shown in Figure 4.17 (a), and the other is the vertical RRAM sandwiched between the pillar electrodes and multi-layer plane electrodes, shown in Figure 4.17 (b). Figure 4.17 (c) shows the cross-section schematic of the vertical RRAM by cutting through one pillar electrode: the RRAM cells are formed at the sidewall of the pillar electrode and in contact with the plane electrodes (highlighted by red dash circle), and there is one cell at each metal layer. The fabrication cost of the first approach using stacked horizontal RRAM is relatively higher because the number of lithography steps increases with the number of the layers, thus the fabrication cost remains high as lithography step is expensive. The second approach using vertical RRAM requires only one critical lithography step to define the pillar electrodes after sequentially depositing multi-layer plane electrodes, making it a more promising approach for reducing the fabrication cost. However, the cost per bit analysis for these two 3D array architecture is not so intuitive. Although the vertical RRAM saves the fabrication cost, its minimal F is not as small as that of the horizontal RRAM, thus it has a lower integration density. This is because the diameter of the pillar electrode is limited by the following factors. Firstly, the aspect ratio of the pillar electrode is limited by the etching process capability of metal/dielectric multilayers, thus the number of the stacked layers is limited. Secondly, the pillar electrode resistance will drastically increase at the nanoscale. As a rough estimation, the vertical RRAM can scale to F=30 nm considering a pillar diameter (~20 nm) plus twice the RRAM oxide thickness (~5 nm). If the horizontal RRAM can scale to F=10 nm with the help of the selector, then 1 layer of horizontal RRAM has the same integration density as 9 layers of vertical RRAM. Further detailed analysis is needed to assess the pros and cons of these two 3D integration approaches.

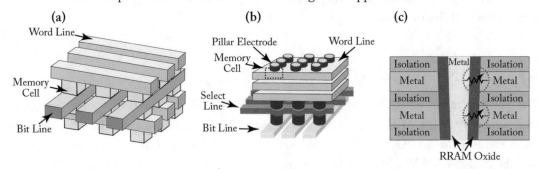

Figure 4.17: Schematic of (a) simply stacked 3D horizontal RRAM array; (b) 3D vertical RRAM array; (c) the cross-section of (b) by cutting through one pillar electrode.

A proof-of-concept two-layer 3D vertical RRAM has been demonstrated in TaO_x [102] and HfO_x [103]. Figure 4.18 shows the fabricated HfO_x-based vertical RRAM [103]. The TEM image in Figure 4.18 (a) shows that the HfO_x switching layer is formed at the side wall between the TiN pillar electrode and the Pt plane electrode. Figure 4.18 (b) shows consistent bipolar switching characteristics between the two-layer vertical RRAM prototype (top cell and bottom cell) and the single-layer control sample. The 3D vertical RRAM cell performances were statistically characterized, and were summarized in Figure 4.18 (c). These experiments have demonstrated that performance of a 3D vertical RRAM at a single bit level shows great promise as a 3D NAND FLASH replacement.

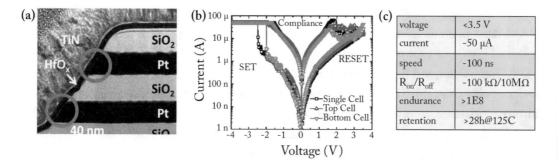

Figure 4.18: (a) Cross-section TEM of the two-layer vertical RRAM prototype. The RRAM cell is formed at the sidewall between TiN pillar electrode and Pt plane electrode. (b) Bipolar switching characteristics of the two-layer vertical RRAM prototype (top cell and bottom cell) and the single-layer control sample. The voltage is referenced to the TiN electrode, which is grounded. (c) Summary of the vertical RRAM prototype performances. Adapted from [103].

Despite the success of the functionality demonstration of the 3D vertical RRAM at the single device level, whether the 3D vertical RRAM can be scaled up to a large array and stacked with multiple layers remains uncertain. There are a few technological challenges at the array level: Firstly, the 3D cross-point array size is still limited by the IR drop and sneak paths, as in the case of the 2D cross-point array discussed in Section 4.3. The 3D cross-point array may have many more sneak paths than the 2D cross-point array, thus the sneak path problem may be exacerbated. In order to suppress the leakage current, a selector device in series with a RRAM cell can be used. However, an external selector is undesired for the 3D vertical RRAM as it inevitably adds the lateral dimension to the cell at the sidewall, since the thickness of the selector adds to the thickness of the cell laterally. Therefore, developing RRAM devices with built-in I-V nonlinearity is of great interest. Secondly, the electrode or interconnect materials for the 3D vertical RRAM is a concern. As shown in the proof-of-concept work of the HfO_x vertical RRAM [103], the vertical pillar is not perfectly vertical due to the fact that Pt is very hard to etch and Pt is not CMOS process compatible. Although TiN is a very common electrode material for RRAM devices, the TiN pillar may introduce significant IR drop on the interconnect as the TiN resistivity is noticeably worse than that of the common metals. Therefore, alternative electrode materials other than TiN or Pt are worth exploration. Thirdly, the drivability of the transistor at the bottom of the pillar electrode may limit the number of layers (N) that can be stacked. To achieve the $4F^2/N$ ultra-high density, the transistor must be a vertical gate-all-around transistor, which also needs to provide sufficient programming current in addition to the sneak path current along the pillar. As a reference, state-of-the-art of vertical silicon transistor can only deliver ~50 μA at 25 nm diameter [104]. Further reducing the RRAM's programming current and suppressing the sneak path current is necessary.

To understand the key parameters that limit the large-scale integration of 3D vertical RRAM cross-point architecture, array macro modeling to assess the write/read margin has been performed [105]. As shown in Figure 4.19 (a), a sub-circuit module with 8 RRAM cells is set up in SPICE simulator, and the sub-circuit module is duplicated to the 3D space to simulate the full 3D cross-point array. Three kinds of resistors are considered: RRAM cell resistance, pillar resistance, and plane resistance. In order to simulate the plane resistance more accurately, a virtual node is added. With this approach, the array design metrics such as write/read margin can be explored as a function of the array geometry dimension, device parameters, and array size. Figure 4.19 (b) and (c) show write access voltage and read sense margin with different on-state resistance (R_{on}) as a function of array planar size for a 16-layer vertical RRAM array. It is seen that increasing R_{on} to above 500 kΩ is an effective way to increase write access voltage while maintaining the read sense margin above the criterion (~100 nA for state-of-the art current-mode sense amplifier [82]). It can be seen that there is a fundamental trade-off between the write access voltage and the read sense margin which limits the upper-bound of the R_{on}. By a careful design of the RRAM device parameters and the voltage settings for write/read, the 3D vertical RRAM has the potential to achieve 1 Mb sub-array (or block) size.

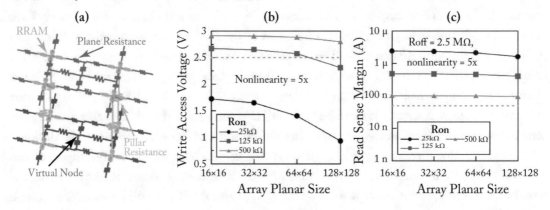

Figure 4.19: (a) Schematic of the sub-circuit module with 8 RRAM cells is set up in SPICE simulator. (b) Write access voltage and (c) read sense margin with different on-state resistance (Ron) as a function of array planar size for a 16-layer vertical RRAM array. Adapted from [105].

CHAPTER 5

Outlook for RRAM's Applications

While RRAM has the potential to be a standalone and large-capacity NVM technology to replace NOR FLASH or NAND FLASH, it may be also suitable for embedded applications. This is because it offers the low programming voltage that embedded FLASH does not offer, and it also offers the non-volatility that DRAM does not offer and yet has a speed that is comparable to DRAM. In this sense, RRAM is attractive to serve as the storage class memory [7] between the DRAM and FLASH in the memory hierarchy. For storage class memory applications, further improving the cycling endurance and lowering the programming voltages to be compatible with the CMOS logic process are of higher priority than improving other attributes. The requirement of the cycling endurance can be estimated as follows: if the application target is to place the RRAM close to the DRAM-based main memory, the I/O transfer rate is typically 1.6 Gb/s. Assuming all the transfer events are write events, for a 10-year lifetime (~3×10^8 s), the endurance of a single cell should be 4.8×10^{17} cycles, which seems impossible for RRAM to achieve. However, with the architectural wear-leveling techniques [106], the write events can be uniformly distributed to different sub-arrays or blocks. Assuming for a 1 Gb RRAM array macro with a 10% wear-leveling effectiveness, the endurance for each cell reduces to 4.8×10^9, which seems reasonable for many RRAM devices to achieve. If the application target is to place the RRAM close to the NAND FLASH, then the requirement of cycling endurance will be much reduced.

Besides the commercial NVM applications, RRAM is also attractive for radiation-hard NVM for the aerospace electronics or other harsh radiation environments. Many experiments show that RRAM is robust against the radiation effects such as total ionizing dose effect [107, 41, 108], while the single-event-upset effect observed in the RRAM was attributed to the photocurrent generated at the neighboring transistor's drain to body p-n junction [41, 109], which can be eliminated by using silicon-on-insulator (SOI) transistors.

Beyond the NVM applications, novel applications for RRAM are emerging. Firstly, the use of RRAM as the reconfigurable switch has been proposed. RRAM-based field programmable gate array (FPGA) was designed [110] and fabricated [111]. Secondly, the use of RRAM as ternary content-addressable-memory (TCAM) for fast searching big data has been reported [112]. Thirdly, the use of RRAM as the physical unclonable function (PUF) for hardware security primitive was also proposed [113, 114], which leverages the intrinsic variations in the RRAM switching processes. Another emerging application is using RRAM as synaptic devices for hardware implementation of neuro-inspired computing [115]. Owing to RRAM's multi-level capability, it serves an analog memory emulating the function of plastic synapses in a neural network, and the cross-point architecture can efficiently implement the weighted sum and weight update process in the learning algorithms [116].

Although the early vision for RRAM is to strive for a 4 F^2 cross-point architecture with multi-level operation and 3D integration capability, it is not entirely clear that these goals continue to make sense generally, given the many diverse potential applications of RRAM. By taking advantages of RRAM, there are enormous opportunities to completely re-think the design of the computer systems to gain orders of magnitude improvement in speed and/or power consumption. RRAM's unique physical properties may also add new functionality and features to the systems. A revolution of future's computing paradigm will bring about a fundamental change in how one can extract benefits out of the technology advancements.

Bibliography

[1] H.-S. P. Wong and S. Salahuddin, "Memory leads the way to better computing," *Nature Nanotechnology*, vol. 10, pp. 191–194, 2015. DOI: 10.1038/nnano.2015.29. 1

[2] B. J. Zhu, "Magnetoresistive random access memory: the path to competitiveness and scalability," *Proceedings of the IEEE*, vol. 96, no. 11, p. 1786–1798, 2008. DOI: h10.1109/JPROC.2008.2004313. 1

[3] H.-S. P. Wong, S. Raoux, S. Kim, J. Liang, J. P. Reifenberg, B. Rajendran, M. Asheghi, and K. E. Goodson, "Phase change memory," *Proceedings of the IEEE*, vol. 98, no. 12, p. 2201–2227, 2010. DOI: 10.1109/JPROC.2010.2070050. 1

[4] H.-S. P. Wong, H.-Y. Lee, S. Yu, Y.-S. Chen, Y. Wu, P.-S. Chen, B. Lee, F. T. Chen, and M.-J. Tsai, "Metal–oxide RRAM," *Proceedings of the IEEE*, vol. 100, no. 6, p. 1951–1970, 2012. DOI: 10.1109/JPROC.2012.2190369. 1, 3, 22

[5] S. P. Park, S. Gupta, N. Mojumder, A. Raghunathan, and K. Roy, "Future cache design using STT MRAMs for improved energy efficiency: devices, circuits and architecture," *ACM Design Automation Conference*, 2012. DOI: 10.1145/2228360.2228447. 2

[6] M. Jung, J. Shalf, and M. Kandemir, "Design of a large-scale storage-class RRAM system," *ACM International Conference on Supercomputing*, 2013. DOI: 10.1145/2464996.2465004. 2

[7] R. F. Freitas and W. W. Wilcke, "Storage-class memory: The next storage system technology," *IBM Journal of Research and Development*, vol. 52, no. 4.5, pp. 439–447, 2008. 2

[8] S. Tanakamaru, H. Yamazawa, T. Tokutomi, S. Ning, and K. Takeuchi, "Hybrid storage of ReRAM/TLC NAND Flash with RAID-5/6 for cloud data centers," *IEEE International Solid-State Circuits Conference*, 2014. DOI: 10.1109/isscc.2014.6757459. 2

[9] T. W. Hickmott, "Low-frequency negative resistance in thin anodic oxide films," *Journal of Applied Physics*, vol. 33, no. 9, p. 2669–2682, 1962. DOI: 10.1063/1.1702530. 3

[10] A. Asamitsu, Y. Tomioka, H. Kuwahara, and Y. Tokura, "Current switching of resistive states in magnetoresistive manganites," *Nature*, vol. 388, no. 3, p. 1995–1997, 1997. 3

[11] A. Beck, J. G. Bednorz, C. Gerber, C. Rossel, and D. Widmer, "Reproducible switching effect in thin oxide films for memory applications," *Applied Physics Letters*, vol. 77, no. 1, p. 139–141, 2000. DOI: 10.1063/1.126902. 3

[12] Y. Watanabe, J. G. Bednorz, A. Bietsch, C. Gerber, D. Widmer, A. Beck, and S. J. Wind, "Current-driven insulator–conductor transition and nonvolatile memory in chromium-doped SrTiO3 single crystals," *Applied Physics Letters*, vol. 78, no. 23, p. 3738–3740, 2001. DOI: 10.1063/1.1377617. 3

[13] I. G. Baek, M. S. Lee, S. Sco, M. J. Lee, D. H. Seo, D.-S. Suh, J. C. Park, S. O. Park, H. S. Kim, I. K. Yoo, U.-I. Chung, and J. T. Moon, "Highly scalable non-volatile resistive memory using simple binary oxide driven by asymmetric unipolar voltage pulses," *IEEE International Electron Devices Meeting*, 2004. DOI: 10.1109/iedm.2004.1419228. 3

[14] S. Seo, M. J. Lee, D. H. Seo, E. J. Jeoung, D.-S. Suh, Y. S. Joung, I. K. Yoo, I. R. Hwang, S. H. Kim, I. S. Byun, J.-S. Kim, J. S. Choi, and B. H. Park, "Reproducible resistance switching in polycrystalline NiO films," *Applied Physics Letters*, vol. 85, no. 23, p. 5655–5657, 2004. DOI: 10.1063/1.1831560. 3

[15] B. J. Choi, D. S. Jeong, S. K. Kim, C. Rohde, S. Choi, J. H. Oh, H. J. Kim, C. S. Hwang, K. Szot, R. Waser, B. Reichenberg, and S. Tiedke, "Resistive switching mechanism of TiO2 thin films grown by atomic-layer deposition," *Journal of Applied Physics*, vol. 98, no. 3, p. 033715, 2005. 3

[16] A. Chen, S. Haddad, Y. J. Wu, T. Fang, Z. Lan, S. Avanzino, M. Buynoski, M. Rathor, W. D. Cai, N. Tripsas, C. Bill, M. Vanbuskirk, and M. Taguchi, "Non-volatile resistive switching for advanced memory applications," *IEEE International Electron Devices Meeting*, 2005. 3

[17] C. Lin, C. Wu, C. Wu, T. Tseng, and C. Hu, "Modified resistive switching behavior of ZrO2 memory films based on the interface layer formed by using Ti top electrode," *Journal of Applied Physics*, vol. 102, no. 9, p. 094101, 2007. DOI: 10.1063/1.2802990. 3

[18] N. Xu, L. Liu, X. Sun, X. Liu, D. Han, Y. Wang, R. Han, J. Kang, and B. Yu, "Characteristics and mechanism of conduction/set process in TiNZnOPt resistance switching random-access memories," *Applied Physics Letters*, vol. 92, no. 23, p. 232112, 2008. DOI: 10.1063/1.2945278. 3

[19] H. Y. Lee, P. S. Chen, T. Y. Wu, Y. S. Chen, C. C. Wang, P. J. Tzeng, C. H. Lin, F. Chen, C. H. Lien, and M. Tsai, "Low power and high speed bipolar switching with a thin reactive Ti buffer layer in robust HfO2 based RRAM," *IEEE International Electron Devices Meeting*, 2008. DOI: 10.1109/iedm.2008.4796677. 3, 5, 6, 13, 14

[20] Z. Wei, Y. Kanzawa, K. Arita, Y. Katoh, K. Kawai, S. Muraoka, S. Mitani, S. Fujii, K. Katayama, M. Iijima, T. Mikawa, T. Ninomiya, R. Miyanaga, Y. Kawashima, K. Tsuji, A. Himeno, T. Okada, R. Azuma, K. Shimakawa, H. Sugaya, T. Takagi, R. Yasuhara, and K.

Hori, "Highly reliable TaOx ReRAM and direct evidence of redox reaction mechanism," *IEEE International Electron Devices Meeting*, 2008. DOI: 10.1109/IEDM.2008.4796676. 3, 7, 19

[21] W. Kim, S. I. Park, Z. Zhang, Y. Y. Liauw, D. Sekar, H.-S. P. Wong, and S. S. Wong, "Forming-free nitrogen-doped AlOx RRAM with sub-μA programming Current," *IEEE Symposium on VLSI Technology*, 2011. 3

[22] D. B. Strukov, G. S. Snider, D. R. Stewart, and R. S. Williams, "The missing memristor found," *Nature*, vol. 453, pp. 80–83, 2008. DOI: 10.1038/nature06932. 3

[23] H. Akinaga and H. Shima, "Resistive random access memory (ReRAM) based on metal oxides," *Proceedings of the IEEE*, vol. 98, no. 12, pp. 2237–2251, 2010. DOI: 10.1109/ JPROC.2010.2070830. 3

[24] R. Waser, R. Dittmann, G. Staikov, and K. Szot,, "Redox-based resistive switching memories - nanoionic mechanisms, prospects, and challenges," *Advanced Materials*, vol. 21, no. 25–26, p. 2632–2663, 2009. DOI: 10.1002/adma.200900375. 3, 21

[25] R. Waser and M. Aono, "Nanoionics-based resistive switching memories," *Nature Materials*, vol. 6, p. 833–840, 2007. DOI:10.1038/nmat2023. 3

[26] I. Valov, R. Waser, J. R. Jameson, and M. N. Kozicki , "Electrochemical metallization memories—fundamentals, applications, prospects," *Nanotechnology*, vol. 22, p. 254003, 2011. DOI: 10.1088/0957-4484/22/25/254003. 3

[27] S. Yu and H.-S. P. Wong, "A phenomenological model for the reset mechanism of metal oxide RRAM," *IEEE Electron Device Letters*, vol. 31, no. 12, p. 1455–1457, 2010. DOI: 10.1109/LED.2010.2078794. 5, 21

[28] B. Govoreanu, G. S. Kar, Y. Chen, V. Paraschiv, S. Kubicek, A. Fantini, I. P. Radu, L. Goux, S. Clima, R. Degraeve, N. Jossart, O. Richard, T. Vandeweyer, K. Seo, P. Hendrickx, G. Pourtois, H. Bender, L. Altimime, D. J. Wouters, J. A. Kittl, and M. Jurczak, "10×10nm2 Hf /HfOx crossbar resistive RAM with excellent performance , reliability and low-energy operation," *IEEE International Electron Devices Meeting*, 2011. DOI: 10.1109/IEDM.2011.6131652. 5, 7, 9, 10, 11, 12, 13

[29] Y. S. Chen, H. Y. Lee, P. S. Chen, P. Y. Gu, C. W. Chen, W. P. Lin, W. H. Liu, Y. Y. Hsu, S. S. Sheu, P. C. Chiang, W. S. Chen, F. T. Chen, C. H. Lien, and M. J. Tsai, "Highly scalable hafnium oxide memory with improvements of resistive distribution and read disturb immunity," *IEEE International Electron Devices Meeting*, 2009. DOI: 10.1109/ IEDM.2009.5424411. 5, 6

[30]   K.-S. Li, C. H. Ho, M.-T. Lee, M.-C. Chen, C.-L. Hsu, J. M. Lu, C. H. Lin, C. C. Chen, B. W. Wu, Y. F. Hou, C. Y. Lin, Y. J. Chen, T. Y. Lai, M. Y. Li, I. Yang, C. S. Wu, and F.-L. Yang, "Utilizing Sub-5 nm sidewall electrode technology for atomic-scale resistive memory fabrication," *IEEE Symposium on VLSI Technology*, 2014. DOI: 10.1109/VLSIT.2014.6894402. 7, 11

[31]   M.-J. Lee, C. B. Lee, D. Lee, S. R. Lee, M. Chang, J. H. Hur, Y.-B. Kim, C.-J. Kim, D. H. Seo, S. Seo, U.-I. Chung, I.-K. Yoo, and K. Kim, "A fast, high-endurance and scalable non-volatile memory device made from asymmetric Ta2O5-x/TaO2-x bilayer structures," *Nature Materials*, vol. 10, p. 625–630, 2011. DOI: 10.1038/nmat3070. 7, 16

[32]   L. Zhao, Z. Jiang, H.-Y. Chen, J. Sohn, K. Okabe, B. Magyari-Kope, H.-S. P. Wong, and Y. Nishi, "Ultrathin (~ 2nm) HfOx as the fundamental resistive switching element: Thickness scaling limit, stack engineering and 3D integration," *IEEE International Electron Devices Meeting*, 2014. DOI: 10.1109/IEDM.2014.7046998. 7

[33]   W.-C. Chien, M.-H. Lee, Y.-Y. Lin, and K.-Y. Hsieh, "Multi-level 40nm WOx resistive memory with excellent reliability," *IEEE International Electron Devices Meeting*, 2011. DOI: 10.1109/IEDM.2011.6131651. 7, 14

[34]   S. R. Lee, Y.-B. Kim, M. Chang, K. M. Kim, C. B. Lee, J. H. Hur, D.-S. Park, D. Lee, M.-J. Lee, C. J. Kim, U. Chung, I.-K. Yoo, and K. Kim, "Multi-level switching of triple-layered TaOx RRAM with excellent reliability for storage class memory," *IEEE Symposium on VLSI Technology*, 2012. DOI: 10.1109/VLSIT.2012.6242466. 7

[35]   S.-S. Sheu, M.-F. Chang, K.-F. Lin, C.-W. Wu, Y.-S. Chen, P.-F. Chiu, C.-C. Kuo, Y.-S. Yang, P.-C. Chiang, W.-P. Lin, C.-H. Lin, H.-Y. Lee, P.-Y. Gu, S.-M. Wang, F. T. Chen, K.-L. Su, C.-H. Lien, K.-H. Cheng, H.-T. Wu, T.-K. Ku, M.-J. Kao, and M.-J. Tsai, "A 4Mb embedded SLC resistive-RAM macro with 7.2ns read-write random-access time and 160ns MLC-access capability," *IEEE International Solid-State Circuits Conference*, 2011. DOI: 10.1109/ISSCC.2011.5746281. 7, 14, 15, 35, 36

[36]   A. Kawahara, R. Azuma, Y. Ikeda, K. Kawai, Y. Katoh, K. Tanabe, T. Nakamura, Y. Sumimoto, N. Yamada, N. Nakai, S. Sakamoto, Y. Hayakawa, K. Tsuji, S. Yoneda, A. Himeno, K. Origasa, K. Shimakawa, T. Takagi, T. Mikawa, and K. Aono, "An 8Mb multi-layered cross-point ReRAM macro with 443MB/s write throughput," *IEEE International Solid-State Circuits Conference*, 2012. DOI: 10.1109/isscc.2012.6177078. 7, 39, 40

[37]   T.-Y. Liu, T. H. Yan, R. Scheuerlein, Y. Chen, J. K. Lee, G. Balakrishnan, G. Yee, H. Zhang, A. Yap, J. Ouyang, T. Sasaki, S. Addepalli, A. Al-Shamma, C.-Y. Chen, M. Gupta, G. Hilton, S. Joshi, A. Kathuria, V. Lai, D. Masiwal, M. Matsumoto, et al. , "A 130.7mm2

2-layer 32Gb ReRAM memory device in 24nm technology," *IEEE International Solid-State Circuits Conference*, 2013. 7, 39, 40, 41

[38]   J. H., Stathis, "Percolation models for gate oxide breakdown," *Journal of Applied Physics*, vol. 86, no. 10, pp. 5757–5766, 1999. DOI: 10.1063/1.371590. 10

[39]   S. Yu, Y. Wu, and H.-S. P. Wong, "Investigating the switching dynamics and multi-level capability of bipolar metal oxide resistive switching memory," *Applied Physics Letters*, vol. 98, p. 103514, 2011. 12

[40]   H. Y. Lee, Y. S. Chen, P. S. Chen, P. Y. Gu, Y. Y. Hsu, S. M. Wang, W. H. Liu, C. H. Tsai, S. S. Sheu, P. C. Chiang, W. P. Lin, C. H. Lin, W. S. Chen, F. T. Chen, C. H. Lien, and M.-J. Tsai, "Evidence and solution of over-RESET problem for HfOx based resistive memory with sub-ns switching speed and high endurance," *IEEE International Electron Devices Meeting*, 2010. 12

[41]   W. G. Bennett, N. C. Hooten, R. D. Schrimpf, R. A. Reed, M. H. Mendenhall, M. L. Alles, J. Bi, E. X. Zhang, D. Linten, M. Jurzak, and A. Fantini, "Single- and multiple-event induced upsets in HfO2/Hf 1T1R RRAM," *IEEE Transactions on Nuclear Science*, vol. 61, no. 4, pp. 1717–1725, 2014. DOI: 10.1109/TNS.2014.2321833. 12, 55

[42]   S. Yu, X. Guan and H.-S. P. Wong, "On the stochastic nature of resistive switching in metal oxide RRAM: physical modeling, Monte Carlo simulation, and experimental characterization," *IEEE International Electron Devices Meeting*, 2011. DOI: 10.1109/iedm.2011.6131572. 13, 28

[43]   Y. Y. Chen, B. Govoreanu, L. Goux, R. Degraeve, Andrea Fantini, G. S. Kar, D. J. Wouters, G. Groeseneken, J. A. Kittl, M. Jurczak, and L. Altimime, "Balancing SET/RESET pulse for > 1E10 endurance in HfO2/Hf 1T1R bipolar RRAM," *IEEE Transactions on Electron Devices*, vol. 59, no. 12, pp. 3243–3249, 2012. DOI: 10.1109/TED.2012.2218607. 15, 17, 18, 33

[44]   Y. Y. Chen, R. Degraeve, S. Clima, B. Govoreanu, L. Goux, A. Fantini, G. S. Kar, G. Pourtois, G. Groeseneken, D. J. Wouters, and M. Jurczak, "Understanding of the endurance failure in scaled HfO2-based 1T1R RRAM through vacancy mobility degradation," *IEEE International Electron Devices Meeting*, 2012. DOI: 10.1109/IEDM.2012.6479079. 18, 19

[45]   Y. Y. Chen, M. Komura, R. Degraeve, B. Govoreanu, L. Goux, A. Fantini, N. Raghavan, S. Clima, L. Zhang, A. Belmonte, A. Redolfi, G. S. Kar, G. Groeseneken, D. J. Wouters, and M. Jurczak, "Improvement of data retention in HfO2 / Hf 1T1R RRAM cell under low

operating current," *IEEE International Electron Devices Meeting*, 2013. DOI: 10.1109/LED.2013.2251857. 19, 33, 34

[46] Y. Y. Chen, R. Degraeve, B. Govoreanu, S. Clima, L. Goux, A. Fantini, G. S. Kar, D. J. Wouters, G. Groeseneken, and M. Jurczak, "Postcycling LRS retention analysis in HfO2/Hf RRAM 1T1R device," *IEEE Electron Device Letters*, vol. 34, no. 5, pp. 636–638, 2013 DOI: 10.1109/LED.2013.2251857. 19

[47] Z. Wei, T. Takagi, Y. Kanzawa, Y. Katoh, T. Ninomiya, K. Kawai, S. Muraoka, S. Mitani, K. Katayama, S. Fujii, R. Miyanaga, Y. Kawashima, T. Mikawa, K. Shimakawa, and K. Aono,, "Demonstration of high-density ReRAM ensuring 10-year retention at 85°C based on a newly developed reliability model," *IEEE International Electron Devices Meeting*, 2011. DOI: 10.1109/iedm.2011.6131650. 19

[48] R. Meyer, L. Schloss, J. Brewer, R. Lambertson, W. Kinney, J. Sanchez, and D. Rinerson, "Oxide dual-layer memory element for scalable non-volatile cross-point memory technology," *IEEE Non-Volatile Memory Technology Symposium*, 2008. DOI: 10.1109/nvmt.2008.4731194. 21

[49] M. Fujimoto, H. Koyama, M. Konagai, Y. Hosoi, K. Ishihara, S. Ohnishi, and N. Awaya,, "TiO2 anatase nanolayer on TiN thin film exhibiting high-speed bipolar resistive switching," *Applied Physics Letters*, vol. 89, no. 22, p. 223509, 2006. DOI: 10.1063/1.2397006. 21

[50] G. Bersuker, D. C. Gilmer, D. Veksler, J. Yum, H. Park, S. Lian, L. Vandelli, A. Padovani, L. Larcher, K. Mckenna, A. Shluger, V. Iglesias, M. Porti, M. Nafria, W. Taylor, P. D. Kirsch, and R. Jammy, "Metal oxide RRAM switching mechanism based on conductive filament microscopic properties," *IEEE International Electron Devices Meeting*, 2010. DOI: 10.1109/iedm.2010.5703394. 21

[51] B. Gao, J. F. Kang, Y. S. Chen, F. F. Zhang, B. Chen, P. Huang, L. F. Liu, X. Y. Liu, Y. Y. Wang, X. A. Tran, Z. R. Wang, H. Y. Yu, and A. Chin, "Oxide-based RRAM: unified microscopic principle for both unipolar and bipolar switching," *IEEE International Electron Devices Meeting*, 2011. DOI: 10.1109/iedm.2011.6131573. 21

[52] U. Celano, Y. Y. Chen, D. J. Wouters, G. Groeseneken, M. Jurczak, and W. Vandervorst, "Filament observation in metal-oxide resistive switching devices," *Applied Physics Letters*, vol. 102, p. 121602, 2013. DOI: 10.1109/iedm.2011.6131573. 22, 23

[53] D.-H. Kwon, K. M. Kim, J. H. Jang, J. M. Jeon, M. H. Lee, G. H. Kim, X.-S. Li, G.-S. Park, B. Lee, S. Han, M. Kim, and C. S. Hwang, "Atomic structure of conducting nanofilaments in TiO2 resistive switching memory," *Nature Nanotechnology*, vol. 5, no. 2, pp. 148–153, 2010. DOI: 10.1038/nnano.2009.456. 23

[54]   X. Cartoixa, R. Rurali, and J. Sune, "Transport properties of oxygen vacancy filaments in metal/crystalline or amorphous HfO2/metal structures," *Physical Review B*, vol. 86, no. 16, p. 165445, 2012. DOI: 10.1103/PhysRevB.86.165445. 23

[55]   P. Calka, E. Martinez, V. Delaye, D. Lafond, G. Audoit, D. Mariolle, N. Chevalier, H. Grampeix, C. Cagli, V. Jousseaume, and C. Guedj, "Chemical and structural properties of conducting nanofilaments in TiN/HfO2-based resistive switching structures," *Nanotechnology*, vol. 24, no. 8, p. 085706, 2013. DOI: 10.1088/0957-4484/24/8/085706. 23

[56]   S. Privitera, G. Bersuker, B. Butcher, A. Kalantarian, S. Lombardo, C. Bongiorno, R. Geer, D. C. Gilmer, and P. D. Kirsch, "Microscopy study of the conductive filament in HfO2 resistive switching memory devices," *Microelectronic Engineering*, Vols. 75–78, p. 109, 2013. DOI: 10.1016/j.mee.2013.03.145. 23, 24

[57]   R. Fang, W. Chen, L. Gao, W. Yu, and S. Yu, "Low temperature characteristics of HfOx-based resistive random access memory," *IEEE Electron Device Letters*, 2015. DOI: 10.1109/LED.2015.2420665. 25

[58]   S. Yu, X. Guan, and H.-S. P. Wong, "Conduction mechanism of TiN/HfOx/Pt resistive switching memory: a trap-assisted-tunneling model," *Applied Physics Letters*, vol. 99, p. 063507, 2011. DOI: 10.1063/1.3624472. 25, 26

[59]   S. Yu, R. Jeyasingh, Y. Wu, and H.-S. P. Wong, "Understanding the conduction and switching mechanism of metal oxide RRAM through low frequency noise and AC conductance measurement and analysis," *IEEE International Electron Devices Meeting*, 2011. DOI: 10.1109/iedm.2011.6131537. 26, 27

[60]   M. Nardone, V. Kozub, I. Karpov, and V. Karpov, "Possible mechanisms for 1/f noise in chalcogenide glasses: a theoretical description," *Physical Review B*, vol. 79, no. 16, p. 165206, 2009. DOI: 10.1103/PhysRevB.79.165206. 26

[61]   S. Balatti, S. Ambrogio, A. Cubeta, A. Calderoni, N. Ramaswamy, and D. Ielmini, "Voltage-dependent random telegraph noise (RTN) in HfOx resistive RAM," *IEEE International Reliability Physics Symposium*, 2014. 27

[62]   D. Veksler, G. Bersuker, L. Vandelli, A. Padovani, L. Larcher, A. Muraviev, B. Chakrabarti, E. Vogel, D. C. Gilmer, and P. D. Kirsch, "Random telegraph noise (RTN) in scaled RRAM devices," *IEEE International Reliability Physics Symposium* , 2013. 27

[63]   S. Yu, X. Guan, and H.-S. P. Wong, "Understanding metal oxide RRAM current overshoot and reliability using Kinetic Monte Carlo simulation," *IEEE International Electron Devices Meeting*, 2012. DOI: 10.1109/iedm.2012.6479105. 28, 29

[64]   A. Padovani, L. Larcher, O. Pirrotta, L. Vandelli, and G. Bersuker, "Microscopic modeling of HfOx RRAM operations: from forming to switching," *IEEE Transcations on Electron Devices*, vol. 62, no. 6, pp. 1998–2006, 2015. DOI: 10.1109/TED.2015.2418114. 28, 29, 30, 31

[65]   S. Larentis, F. Nardi, S. Balatti, D. C. Gilmer, and D. Ielmini, "Resistive switching by voltage-driven ion migration in bipolar RRAM—Part II: Modeling," *IEEE Transactions on Electron Devices*, vol. 59, no. 9, pp. 2468–2475, 2012. DOI: 10.1109/TED.2012.2202320. 31

[66]   S. Kim, S.-J. Kim, K. M. Kim, S. R. Lee, M. Chang, E. Cho, Y.-B. Kim, C. J. Kim, U.-I. Chung, and I.-K. Yoo, "Physical electro-thermal model of resistive switching in bi-layered resistance-change memory," *Scientifc Reports*, vol. 3, p. 1680, 2013. DOI: 10.1038/srep01680. 31

[67]   P. Huang, X. Y. Liu, B. Chen, H. T. Li, Y. J. Wang, Y. X. Deng, K. L. Wei, L. Zeng, B. Gao, and G. Du, "A physics-based compact model of metal-oxide-based RRAM DC and AC operations," *IEEE Transactions on Electron Devices*, vol. 60, no. 12, pp. 4090–4097, 2013. DOI: 10.1109/TED.2013.2287755. 31

[68]   M. Bocquet, D. Deleruyelle, H. Aziza, C. Muller, J.-M. Portal, T. Cabout, and E. Jalaguier, "Robust compact model for bipolar oxide-based resistive switching memories," *IEEE Transactions on Electron Devices*, vol. 61, no. 3, pp. 674–681, 2014. DOI: 10.1109/TED.2013.2296793. 31

[69]   L. Larcher, F. M. Puglisi, P. Pavan, A. Padovani, L. Vandelli, and G. Bersuker, "A compact model of program window in HfOx RRAM devices for conductive filament characteristics analysis," IEEE Transactions on Electron Devices, vol. 61, no. 8, pp. 2668–2673, 2014. DOI: 10.1109/TED.2014.2329020. 31

[70]   X. Guan, S. Yu, and H.-S. P. Wong, "A SPICE compact model of metal oxide resistive switching memory with variations," *IEEE Electron Device Letters*, vol. 10, no. 1405–1407, p. 33, 2012. DOI: 10.1109/led.2012.2210856. 31

[71]   "ASU RRAM model," [Online]. Available: http://faculty.engineering.asu.edu/shimengyu/model-downloads/. 31, 34

[72]   J. McPherson, J.-Y. Kim, A. Shanware, and H. Mogul, "Thermochemical description of dielectric breakdown in high dielectric constant materials," *Applied Physics Letters*, vol. 82, no. 13, pp. 2121–2123, 2003. DOI: 10.1063/1.1565180. 33

[73]   S. Yu, B. Gao, Z. Fang, H. Y. Yu, J. F. Kang, and H.-S. P. Wong, "A neuromorphic visual system using RRAM synaptic devices with sub-pJ energy and tolerance to variability:

experimental characterization and large-scale modeling," *IEEE International Electron Devices Meeting*, 2012. DOI: 10.1109/iedm.2012.6479018. 33

[74]   "Predictive Technology Model (PTM)," [Online]. Available: http://ptm.asu.edu/. 33, 38

[75]   M. Mao, Y. Cao, S. Yu, and C. Chakrabarti, "Optimizing latency, energy, and reliability of 1T1R ReRAM through appropriate voltage settings," *IEEE International Conference on Computer Design*, 2015. 37

[76]   "NVSim," [Online]. Available: http://nvsim.org/. 39

[77]   S. Zuloaga, R. Liu, P.-Y. Chen, and S. Yu, "Scaling 2-layer RRAM cross-point array toward 10 nm node: a device-circuit co-design," *IEEE International Symposium on Circuits and Systems*, 2015. DOI: 10.1109/iscas.2015.7168603. 39

[78]   D. Lee, I. J. Chang, S.-Y. Yoon, J. Jang, D.-S. Jang, W.-G. Hahn, J.-Y. Park, D.-G. Kim, C. Yoon, B.-S. Lim, B.-J. Min, S.-W. Yun, J.-S. Lee, I.-H. Park, K.-R. Kim, J.-Y. Yun, Y. Kim, Y.-S. Cho, K.-M. Kang, S.-H. Joo, J.-Y. Chun, J.-N. Im, et al. , "A 64Gb 533Mb/s DDR interface MLC NAND Flash in sub-20nm technology," *IEEE International Solid-State Circuits Conference*, 2012. DOI: 10.1109/ISSCC.2012.6177077. 39

[79]   J. Liang and H.-S. P. Wong, "Cross-point memory array without cell selectors—Device characteristics and data storage pattern dependencies," *IEEE Transactions on Electron Devices*, vol. 57, no. 10, pp. 2531–2538, 2010. DOI: 10.1109/TED.2010.2062187. 44

[80]   Y. Deng, P. Huang, B. Chen, X. Yang, B. Gao, J. Wang, L. Zeng, G. Du, J. Kang, and X. Liu, "ReRAM crossbar array with cell selection device: a device and circuit interaction study," *IEEE Transactions on Electron Devices*, vol. 60, no. 2, pp. 719–726, 2013. DOI: 10.1109/TED.2012.2231683. 44

[81]   D. Niu, C. Xu, N. Muralimanohar, N. P. Jouppi, and Y. Xie, "Design trade-offs for high density cross-point resistive memory," *ACM/IEEE International Symposium on Low power Electronics and Design*, 2012. DOI: 10.1145/2333660.2333712. 44

[82]   M.-F. Chang, S.-J. Shen, C.-C. Liu, C.-W. Wu, Y.-F. Lin, Y.-C. King, C.-J. Lin, H.-J. Liao, Y.-D. Chih, and H. Yamauchi, "An Offset-Tolerant Fast-Random-Read Current-Sampling-Based Sense Amplifier for Small-Cell-Current Nonvolatile Memory," *IEEE Journal of Solid-State Circuits*, vol. 48, no. 3, pp. 864–877, 2013. DOI: 10.1109/JSSC.2012.2235013. 44

[83]   G. W. Burr, R. S. Shenoy, K. Virwani, P. Narayanan, A. Padilla, B. Kurdi, and H. Hwang, "Access devices for 3D crosspoint memory," *Journal of Vacuum Science & Technology B*, vol. 32, no. 4, p. 40802, 2014. DOI: 10.1116/1.4889999. 44

[84]   M.-J. Lee, Y. Park, B.-S. Kang, S.-E. Ahn, C. Lee, K. Kim, W. Xianyu, G. Stefanovich, J.-H. Lee, S.-J. Chung, Y.-H. Kim, C.-S. Lee, J.-B. Park, I.-G. Baek, and I.-K. Yoo, "2-stack 1D-1R cross-point structure with oxide diodes as switch elements for high density resistance RAM applications," *IEEE International Electron Devices Meeting*, 2007. DOI: 10.1109/iedm.2007.4419061. 45

[85]   G. H. Kim, J. H. Lee, Y. Ahn, W. Jeon, S. J. Song, J. Y. Seok, J. H. Yoon, K. J. Yoon, T. J. Park, and C. S. Hwang, "32×32 crossbar array resistive memory composed of a stacked Schottky diode and unipolar resistive memory," *Advanced Functional Materials*, vol. 23, no. 11, pp. 1440–1449, 2013. DOI: 10.1002/adfm.201202170. 45

[86]   J.-J. Huang, Y.-M. Tseng, W.-C. Luo, C.-W. Hsu, and T.-H. Hou, "One selector-one resistor (1S1R) crossbar array for high-density flexible memory applications," *IEEE International Electron Devices Meeting*, 2011. 45

[87]   W. Lee, J. Park, J. Shin, J. Woo, S. Kim, G. Choi, S. Jung, S. Park, D. Lee, E. Cha, H. D. Lee, S. G. Kim, S. Chung, and H. Hwang, "Varistor-type bidirectional switch (JMAX>107A/cm2, selectivity~104) for 3D bipolar resistive memory arrays," *IEEE Symposium on VLSI Technology*, 2012. 45

[88]   K. Gopalakrishnan, R. S. Shenoy, C. T. Rettner, K. Virwani, D. S. Bethune, R. M. Shelby, G. W. Burr, A. Kellock, R. S. King, K. Nguyen, A. N. Bowers, M. Jurich, B. Jackson, A. M. Friz, T. Topuria, P. M. Rice, and B. N. Kurdi, "Highly-scalable novel access device based on mixed ionic electronic conduction (MIEC) materials for high density phase change memory (PCM) arrays," *IEEE Symposium on VLSI Technology*, 2010. DOI: 10.1109/VLSIT.2010.5556229. 45

[89]   G. W. Burr, K. Virwani, R. S. Shenoy, G. Fraczak, C. T. Rettner, A. Padilla, R. S. King, K. Nguyen, A. N. Bowers, M. Jurich, M. BrightSky, E. A. Joseph, A. J. Kellock, N. Arellano, B. N. Kurdi, K. Gopalakrishnan, "Recovery dynamics and fast (sub-50ns) read operation with access devices for 3D crosspoint memory based on mixed-ionic-electronic-conduction (MIEC)," *IEEE Symposium onVLSI Technology*, 2013. 45

[90]   K. Virwani, G. W. Burr, R. S. Shenoy, C. T. Rettner, A. Padilla, T. Topuria, P. M. Rice, G. Ho, R. S. King, K. Nguyen, A. N. Bowers, M. Jurich, M. BrightSky, E. A. Joseph, A. J. Kellock, N. Arellano, B. N. Kurdi and K. Gopalakrishnan, "Sub-30nm scaling and high-speed operation of fully-confined access-devices for 3D crosspoint memory based on Mixed-Ionic-Electronic-Conduction (MIEC) Materials," *IEEE International Electron Devices Meeting*, 2012. DOI: 10.1109/IEDM.2012.6478967. 45

[91]   M.-J. Lee, Y. Park, D.-S. Suh, E.-H. Lee, S. Seo, D.-C. Kim, R. Jung, B.-S. Kang, S.-E. Ahn, C. B. Lee, D. H. Seo, Y.-K. Cha, I.-K. Yoo, J.-S. Kim, and B. H. Park, "Two series

oxide resistors applicable to high speed and high density nonvolatile memory," *Advanced Materials*, vol. 19, no. 22, pp. 3919–3923, 2007. DOI: 10.1002/adma.200700251. 45

[92]   M. Son, J. Lee, J. Park, J. Shin, G. Choi, S. Jung, W. Lee, S. Kim, S. Park, and H. Hwang, "Excellent selector characteristics of nanoscale VO2 for high-density bipolar ReRAM applications," *IEEE Electron Device Letters*, vol. 32, no. 11, pp. 1579–1581, 2011. DOI: 10.1109/LED.2011.2163697. 45

[93]   C. Ko and S. Ramanathan, "Observation of electric field-assisted phase transition in thin film vanadium oxide in a metal-oxide-semiconductor device geometry," *Applied Physics Letters*, vol. 93, no. 25, p. 252101, 2008. 45

[94]   E. Cha, J. Woo, D. Lee, S. Lee, J. Song, Y. Koo, J. H. Lee, C. G. Park, M. Y. Yang, K. Kamiya, K. Shiraishi, B. Magyari-Kope, Y. Nishi, and H. Hwang, "Nanoscale (~10nm) 3D vertical ReRAM and NbO2 threshold selector with TiN electrode," in *IEEE International Electron Devices Meeting*, 2014. DOI: 10.1109/IEDM.2013.6724602. 45

[95]   S. H. Jo, T. Kumar, S. Narayanan, W. D. Lu, and H. Nazarian, "3D-stackable Crossbar Resistive Memory based on Field Assisted Superlinear Threshold (FAST) Selector," *IEEE International Electron Devices Meeting*, 2014. DOI: 10.1109/IEDM.2014.7046999. 46

[96]   M.-F. Chang, A. Lee, P.-C. Chen, C. J. Lin, Y.-C. King, S.-S. Sheu, and T.-K. Ku, "Challenges and circuit techniques for energy-efficient on-chip nonvolatile memory using memristive devices," *IEEE Journal of Emerging and Selected Topics in Circuits and Systems*, vol. 5, no. 2, pp. 183–193, 2015. DOI: 10.1109/JETCAS.2015.2426531. 48, 49, 50, 51

[97]   H. Tanaka, M. Kido, K. Yahashi, M. Oomura, R. Katsumata, M. Kito, Y. Fukuzumi, M. Sato, Y. Nagata, Y. Matsuoka, Y. Iwata, H. Aochi, and A. Nitayama, "Bit cost scalable technology with punch and plug process for ultra high density flash memory," *IEEE Symposium on VLSI Technology*, 2007. DOI: 10.1109/vlsit.2007.4339708. 51

[98]   J. Jang, H. S. Kim, W. Cho, H. Cho, J. Kim, S. I. Shim, Y. Jang, J. H. Jeong, B. K. Son, D. W. Kim, K. Kim, J. J. Shim, J. S. Lim, K. H. Kim, S. Y. Yi, J. Y. Lim, D. Chung, H. C. Moon, S. Hwang, J. W. Lee, Y. H. Son, U. I. Chung, and W. S. Lee, "Vertical cell array using TCAT (Terabit Cell Array Transistor) technology for ultra high density NAND flash memory," *IEEE Symposium on VLSI Technology*, 2009. 51

[99]   K.-T. Park, J.-m. Han, D. Kim, S. Nam, K. Choi, M.-Su Kim, P. Kwak, D. Lee, Y.-H. Choi, K.-M. Kang, M.-H. Choi, D.-H. Kwak, H.-w. Park, S.-w. Shim, H.-J. Yoon, D. Kim, S.-w. Park, K. Lee, K. Ko, D.-K. Shim, Y.-L. Ahn, J. Park, J. Ryu, D. Kim, et al., "Three-dimensional 128Gb MLC vertical NAND Flash memory with 24-WL stacked

layers and 50MB/s high-speed programming," *IEEE International Solid–State Circuits Conference*, 2014. DOI: 10.1109/ISSCC.2014.6757458. 51

[100]   J.-W. Im, W.-P. Jeong, D.-H. Kim, S.-W. Nam, D.-K. Shim, M.-H. Choi, H.-J. Yoon, D.-H. Kim, Y.-S. Kim, H.-W. Park, D.-H. Kwak, S.-W. Park, S.-M. Yoon, W.-G. Hahn, J.-H. Ryu, S.-W. Shim, K.-T. Kang, S.-H. Choi, J.-D. Ihm, Y.-S. Min, I.-M. Kim, et al., "A 128Gb 3b/cell V-NAND flash memory with 1Gb/s I/O rate," *IEEE International Solid–State Circuits Conference*, 2015. DOI: 10.1109/issc.2015.7062960. 51

[101]   H. S. Yoon, I. G. Baek, J. Zhao, H. Sim, M. Y. Park, H. Lee, G. H. Oh, J. C. Shin, I. S. Yeo, and U. I. Chung,, "Vertical cross-point resistance change memory for ultra-high density non-volatile memory applications," *IEEE Symposium on VLSI Technology*, 2009. 51

[102]   I. G. Baek, C. J. Park, H. Ju, D. J. Seong, H. S. Ahn, J. H. Kim, M. K. Yang, S. H. Song, E. M. Kim, S. O. Park, C. H. Park, C. W. Song, G. T. Jeong, S. Choi, H. K. Kang, and C. Chung, "Realization of vertical resistive memory (VRRAM) using cost effective 3D process," *IEEE International Electron Devices Meeting*, 2011. DOI: 10.1109/IEDM.2011.6131654. 52

[103]   H.-Y. Chen, S. Yu, B. Gao, P. Huang, J. F. Kang, and H.-S. P. Wong, "HfOx based vertical RRAM for cost-effective 3D cross-point architecture without cell selector," *IEEE International Electron Devices Meeting*, 2012. DOI: 10.1109/IEDM.2012.6479083. 52, 53

[104]   B. Yang, K. D. Buddharaju, S. H. G. Teo, N. Singh, G. Q. Lo, and D. L. Kwong, "Vertical silicon-nanowire formation and gate-all-around MOSFET," *IEEE Electron Device Letters*, vol. 29, no. 7, pp. 791–793, 2008. DOI: 10.1109/LED.2008.2000617. 53

[105]   P.-Y. Chen, and S. Yu, "Impact of vertical RRAM device characteristics on 3D cross-point array design," *IEEE International Memory Workshop*, 2014. DOI: 10.1109/imw.2014.6849382. 54

[106]   M. K. Qureshi, M. Franceschini, V. Srinivasan, L. Lastras, B. Abali, and J. Karidis, "Enhancing lifetime and security of PCM-based main memory with start-gap wear leveling," *IEEE/ACM International Symposium on Microarchitecture*, 2009. DOI: 10.1145/1669112.1669117. 55

[107]   X. He, W. Wang, B. Butcher, S. Tanachutiwat, and R. E. Geer, "Superior TID hardness in TiN/HfO2/TiN ReRAMs after proton radiation," *IEEE Transactions on Nuclear Science*, vol. 59, no. 5, pp. 2550–2555, 2012. DOI: 10.1109/TNS.2012.2208480. 55

[108]   R. Fang, Y. Gonzalez-Velo, W. Chen, K. Holbert, M. Kozicki, H. Barnaby, and S. Yu, "Total ionizing dose effect of γ-ray radiation on the switching characteristics and filament

stability of HfOx resistive random access memory," *Applied Physics Letters*, vol. 104, p. 183507, 2014. DOI: 10.1063/1.4875748. 55

[109] C. Dakai, K. Hak, A. Phan, E. Wilcox, K. LaBel, S. Buchner, A. Khachatrian, and N. Roche, "Single-event effect performance of a commercial embedded ReRAM," *IEEE Transactions on Nuclear Science*, vol. 61, no. 6, pp. 3088–3094, 2014. DOI: 10.1109/TNS.2014.2361488. 55

[110] S. Tanachutiwat, M. Liu, and W. Wang, "FPGA based on integration of CMOS and RRAM," *IEEE Transactions on Very Large Scale Integrated Systems*, vol. 19, no. 11, p. 2023–2032, 2011. DOI: 10.1109/TVLSI.2010.2063444. 55

[111] Y. Y. Liauw, Z. Zhang, W. Kim, A. El Gamal, and S. S. Wong,, "Nonvolatile 3D-FPGA with monolithically stacked RRAM-based configuration memory," *IEEE International Solid-State Circuits Conference*, 2012. DOI: 10.1109/ISSCC.2012.6177067. 55

[112] M.-F. Chang, C.-C. Lin, A. Lee, C.-C. Kuo, G.-H. Yang, H.-J. Tsai, T.-F. Chen, S.-S. Sheu, P.-L. Tseng, H.-Y. Lee, and T.-K. Ku, "A 3T1R nonvolatile TCAM using MLC ReRAM with Sub-1ns search time," *IEEE International Solid-State Circuits Conference*, 2015. DOI: 10.1109/ISSCC.2015.7063054. 55

[113] P.-Y. Chen, R. Fang, R. Liu, C. Chakrabarti, Y. Cao, and S. Yu, "Exploiting resistive cross-point array for compact design of physical unclonable function," *IEEE International Symposium on Hardware-Oriented Security and Trust*, 2015. DOI: 10.1109/HST.2015.7140231. 55

[114] A. Chen, "Utilizing the variability of resistive random access memory to implement reconfigurable physical unclonable functions," *IEEE Electron Device Letters*, vol. 36, no. 2, pp. 138–140, 2015. DOI: 10.1109/LED.2014.2385870. 55

[115] D. Kuzum, S. Yu, and H.-S. P. Wong, "Synaptic electronics: materials, devices and applications," *Nanotechnology*, vol. 24, p. 382001, 2013. DOI: 10.1088/0957-4484/24/38/382001. 55

[116] S. Yu, P.-Y. Chen, Y. Cao, L. Xia, Y. Wang, and H. Wu, "Scaling-up resistive synaptic arrays for neuro-inspired architecture: Challenges and prospect," *IEEE International Electron Devices Meeting*, 2015. 56

Author Biography

Shimeng Yu received his B.S. degree in microelectronics from Peking University, Beijing, China, in 2009, and his M.S. degree and Ph.D. degree in electrical engineering from Stanford University, Stanford, CA, in 2011, and 2013, respectively. He did summer internships in IMEC, Belgium, in 2011, and IBM TJ Watson Research Center in 2012. He is currently an assistant professor of electrical engineering and computer engineering at Arizona State University, Tempe, AZ U.S.

His research interests are emerging nano-devices and circuits with focus on the resistive switching memories, and new computing paradigms with focus on the neuro-inspired computing. He has published over 40 journal papers and over 80 conference papers with citations of 2500 and H-index 25 by 2015.

He was awarded the Stanford Graduate Fellowship from 2009–2012, the IEEE Electron Devices Society Masters Student Fellowship in 2010, the IEEE Electron Devices Society Ph.D. Student Fellowship in 2012, the DoD DTRA Young Investigator Award in 2015, and the NSF CAREER Award in 2016. He has been serving on the Technical Committee of Nanoelectronics and Gigascale Systems, IEEE Circuits and Systems Society since 2014.

Printed in the United States
by Baker & Taylor Publisher Services